SPICE
Practical Device Modeling

Other Reference Books of Interest by McGraw-Hill

Handbooks

ANTOGNETTI AND MASSOBRIO • *Semiconductor Device Modeling with SPICE*

BAKER • *C Mathematical Function Handbook*

BENSON • *Audio Engineering Handbook*

BENSON • *Television Engineering Handbook*

CHEN • *Computer Engineering Handbook*

COOMBS • *Printed Circuits Handbook*

DI GIACOMO • *Digital Bus Handbook*

DI GIACOMO • *VLSI Handbook*

FINK AND CHRISTIANSEN • *Electronics Engineers' Handbook*

HARPER • *Electronic Packaging and Interconnection Handbook*

HICKS • *Standard Handbook of Engineering Calculations*

INGLIS • *Electronic Communications Handbook*

JURAN AND GRYNA • *Juran's Quality Control Handbook*

KAUFMAN AND SEIDMAN • *Handbook of Electronics Calculations*

RORABAUGH • *Digital Filter Designer's Handbook*

TUMA • *Engineering Mathematics Handbook*

WAYNANT • *Electro-Optics Handbook*

WILLIAMS AND TAYLOR • *Electronic Filter Design Handbook*

Other

ANTOGNETTI • *Power Integrated Circuits*

BEST • *Phase-Locked Loops*

BUCHANAN • *CMOS/TTL Digital Systems Design*

BUCHANAN • *BiCMOS/CMOS Systems Design*

BYERS • *Printed Circuit Board Design with Microcomputers*

ELLIOTT • *Integrated Circuits Fabrication Technology*

HECHT • *The Laser Guidebook*

KIELKOWSKI • *Inside SPICE*

MUN • *GaAs Integrated Circuits*

SILICONIX • *Designing with Field-Effect Transistors*

SZE • *VLSI Technology*

TSUI • *LSI/VLSI Testability Design*

WATERS • *Active Filter Design*

WOBSCHALL • *Circuit Design for Electronic Instrumentation*

WYATT • *Electro-Optical System Design*

SPICE Practical Device Modeling

Ron M. Kielkowski

McGraw-Hill, Inc.

New York San Francisco Washington, D.C. Auckland Bogotá
Caracas Lisbon London Madrid Mexico City Milan
Montreal New Delhi San Juan Singapore
Sydney Tokyo Toronto

Library of Congress Cataloging-in-Publication Data

Kielkowski, Ron M.
SPICE / practical device modeling / Ron M. Kielkowski.
p. cm.
Includes index.
ISBN 0-07-911524-1
1. Electric apparatus and appliances—Computer simulation.
2. SPICE (Computer file) I. Title.
TK7870.K475 1995
621.3815′2—dc20 95-10985
CIP

1 2 3 4 5 6 7 8 9 0 DOH/DOH 9 0 0 9 8 7 6 5

P/N 034626-7
PART OF
ISBN 0-07-911524-1

The sponsoring editor for this book was Stephen S. Chapman, the editing supervisor was Caroline R. Levine, and the production supervisor was Donald F. Schmidt. This book was set in Century Schoolbook by Donald A. Feldman of McGraw-Hill's Professional Book Group composition unit.

Printed and bound by R. R. Donnelley & Sons Company.

This book is printed on acid-free paper containing a minimum of 10% postconsumer waste.

To my father-in-law,
Maurice Earl Johnson

Contents

Preface

Although SPICE has been with us since 1971, use of the program during the past 10 years has exploded because of the personal computer. IBM's introduction of the desktop personal computer and the widespread availability of SPICE, both in the academic and commercial forms, have made analog circuit simulation available to almost every circuit designer.

But before a designer can fully utilize SPICE, accurate device models must be available. Without device models of resistors, capacitors, inductors, diodes, BJTs, JFETs, MOSFETs, and power MOSFETs, the simulator simply cannot produce accurate results and is of little use to the designer.

The widespread acceptance of SPICE and the proliferation of SPICE-like circuit simulators have led more and more component vendors to offer SPICE models of their own components. But the majority of component vendors still do not offer SPICE models.

Other authors have written on the topic of device modeling with SPICE, but these other texts focus on the mathematics and physics of modeling and have failed to produce a clear, step-by-step procedure to producing device models; they also fail to acknowledge the need for accurate device models of resistors, capacitors, inductors, zener diodes, and power MOSFETs. This text rectifies that situation.

Throughout this text, a practical approach to device modeling has been adopted. The goal of this text is to show readers how to produce device models which have an acceptable amount of accuracy (less than 5 percent rms error between the simulated model and the actual measured device characteristics) in less than one eight-hour day.

To obtain accurate models in a reasonable amount of time, several utility programs have been developed to help extract model parameters from measured I-V and C-V characteristics. The utility programs are found on the disk which accompanies this text. Anyone interested in device modeling will find these programs indispensible.

Also included on the disk are the RSPICE and RGRAPH programs. RSPICE is a PC version of SPICE2g.6. RGRAPH is a graphical post-processor for graphing and plotting RSPICE simulation results.

Chapter Organization

Chapter 1 is an introduction to the book and introduces the concept of a *practical* device model. This chapter introduces the idea that, to be useful, a device model must be accurate and readily available. Chapter 1 sets some guidelines for an accurate model and sets limits on the amount of time and energy we will devote to the pursuit of an accurate model.

Chapter 2 introduces the idea of modeling resistors, capacitors, and inductors. While SPICE was originally developed for integrated circuit designers, today SPICE is used to simulate almost every type of electronic circuit. Most designers realize that many of our electronic circuits are limited by second-order effects of our passive components. Electrical series resistance (ESR), lead inductance, and winding capacitance all influence our electronic circuits. Until now, these second-order effects were ignored in simulation. Chapter 2 presents macromodels (subcircuits) of resistors, capacitors, and inductors which include the effects of parasitic capacitance, lead inductance, electrical series resistance, coil resistance, interwinding capacitance, and temperature. The models are described in detail, and step-by-step extraction procedures are presented with each model.

Chapter 3 introduces readers to modeling rectifier and zener diodes. In this chapter, the standard SPICE diode model parameter and equations are explained and examined. Step-by-step procedures demonstrate how to extract model parameter values for the SPICE diode.

Zener diodes cannot be accurately modeled with the standard SPICE diode equations. Chapter 3 continues by introducing a zener diode macromodel which accurately models reverse-bias, temperature-dependent breakdown. Extraction procedures illustrate how to calculate parameter values for the zener model. The chapter closes by modeling a 1n752 zener diode.

Chapter 4 introduces the reader to modeling the bipolar junction transistor (BJT). The model parameters of the SPICE BJT are examined in detail. Step-by-step extraction procedures illustrate how to extract each parameter of the transistor. In this chapter, a model of a 2n2222a bipolar transistor is demonstrated.

Chapter 5 introduces the reader to modeling the junction field-effect transistor (JFET). The model parameters of the SPICE JFET are examined and step-by-step procedures illustrate how each parameter should be extracted. To demonstrate the extraction procedures, a 2n4416 JFET is modeled.

Chapter 6 introduces the reader to modeling the power metal-oxide-semiconductor field-effect transistor (power MOSFET). While the SPICE MOSFET model is suitable for modeling small-signal (lateral) transistors, the power MOSFET and other types of vertical MOSFET transistors cannot accurately be simulated with this model. To circumvent this problem, a power MOSFET macromodel is introduced. In this chapter, step-by-step extraction procedures illustrate how to extract model parameters for the power MOSFET. To demonstrate the extraction procedures, a MTD5N06 power MOSFET is modeled.

Appendixes A through C and App. E are user's guides to modeling the utility programs included on the disk. The appendixes discuss the TC_CALC temperature calculator, CJ_CALC junction capacitance calculator, LIN_REG linear regression estimator, and RS_CALC series resistance calculator. These utility programs are indispensible in modeling devices for SPICE.

Appendix D discusses analog behavioral modeling. Behavioral modeling is modeling in terms of mathematical relationships rather than electrical relationships. The behavioral models in this section includes ideal adders, multipliers, integrators, etc. With these mathematical functions, we can improve the accuracy of many of our standard SPICE models. In many cases we can even develop models not found in SPICE.

Ron M. Kielkowski

Acknowledgments

I would like to thank the many people who contributed to the development of this book. I would like to begin by thanking Art Mernone for reading the manuscript for clarity and correctness; Michael Small and Larry Eves for letting me use their measurement equipment to extract model parameters; Steve Chapman, my editor at McGraw-Hill, for being the kind of editor writers like to write for, and of course, Ronald Rohrer, Donald Pederson, Larry Nagel, and the other Berkeley students and professors who contributed to the development of SPICE and whose efforts changed the face of analog design.

Finally, I would like to thank my family and my beautiful wife Christine, who for the second time in three years, put her dreams on hold so that I could pursue mine.

Chapter

1

Practical Device Modeling

Introduction

As circuit complexity increases and design cycle time decreases, computer simulation of electronic circuits plays an increasingly important role in the design of electronics. And of the many simulation tools available to designers, there is probably none more important than the general-purpose analog simulator SPICE. Universities have integrated SPICE simulations into their engineering curriculum so young engineers can observe the benefits of computer analysis early in their careers. Many companies use computer simulation to help reduce the number of design cycles a product goes through before being brought to market. And many of the engineering communities have gone as far as to implement design procedures which integrate circuit simulation in the design flow and make simulation an integral part of the design methodology taught to new engineers. These design teams recognize that, to compete in today's global electronics marketplace, circuit simulation is the key to reducing design cycle time and time to market.

But to achieve this lofty goal of reducing cycle time, simulations must be fast, they must converge, and, above all, they must be accurate. To a lesser extent, the three virtues of circuit simulation (speed, the ability to converge, and accuracy) will be determined by the simulator's control settings (the .OPTIONS statement). But to a much greater extent, these same virtues reflect the quality of the device models used during simulation. The single most important factor in accurate circuit simulation is the use of accurate device models—and not just semiconductor models. If the simulator is to produce useful, meaningful, realistic results, capacitors, inductors, transformers, power devices, and interconnects must be modeled accurately.

Think of the quality of device models as being analogous to the quality of a baker's recipe ingredients. If all of the ingredients are of high quality, the resulting cake will reflect this. But if a rotten egg is used or a cup of salt is used in place of a cup of sugar, the resulting cake will be awful—*no matter how good the rest of the ingredients are!* Likewise, just one poor device model is enough to spoil an otherwise accurate simulation.

So what we need are absolutely accurate device models for simulation, right? Well, not quite. Again, think of the baker who carefully measures all of the ingredients of the recipe. If the baker follows the recipe carefully, measuring and using the exact amount of ingredients, the resulting cake will be wonderful (depending on your taste in cakes). But what if the baker had some errands to run on this day and, instead of baking the cake himself, asked his apprentice to bake the cake? The apprentice, not having the years of experience the baker has, uses a little too much flour and salt, and maybe not quite enough butter or sugar. If the amount of ingredient variation is small, maybe 5 percent or less, chances are nobody will notice any difference in the resulting dessert. But what if the apprentice had tickets to the Pearl Jam concert that day and, instead of taking the time required to carefully measure all the ingredients, the apprentice haphazardly threw the ingredients together, using twice as many eggs as or only half the sugar prescribed in the recipe? The cake would be awful. If the amount of ingredient is off by 50 percent or more, the quality of the resulting pastry will suffer. The same is true for circuit simulation.

Most passive components have a part-to-part variation of at least 1 to 5 percent, and active components usually vary by even more than this. Just as with our baker, most simulation users will probably not notice a significant difference between a simulation done with models which were accurate to 1 percent and a simulation with models which were accurate to 5 percent, yet the time required to develop the more accurate models could be hours or days longer than that required to develop the less accurate ones.

Achieving absolute accuracy can require exhaustive amounts of time. Figure 1.1 illustrates the logarithmic relationship between the accuracy of a device model and the amount of time and effort required to achieve that accuracy. An engineer might spend one hour modeling a power MOSFET, and, by the end of that hour, the model might be accurate to within 20 percent rms error of the actual device. The engineer might need the remainder of the day to achieve a model with only 5 percent rms error. If the engineer is stubborn and tries to achieve 1 percent rms error, he may spend several days obtaining his final goal.

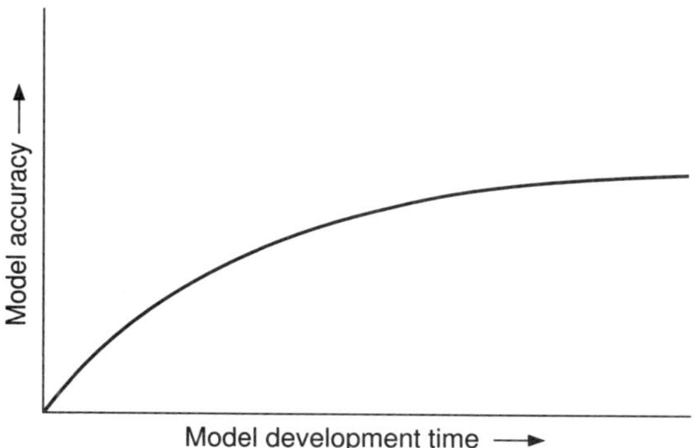

Figure 1.1 Model development time vs. accuracy.

So how much accuracy is required for accurate simulation? In SPICE and the SPICE-like simulators, the accuracy of the simulation engine, *assuming the models are completely accurate,* will be determined by the error tolerances: RELTOL, ABSTOL, and VNTOL. While these can be adjusted, the default setting of them will usually produce an error on the order of less than 1 percent, again assuming completely accurate models. In almost any simulation, the accuracy of the analysis will be limited by the accuracy of the device models rather than by the error tolerances of the program. So again, how much accuracy is required for accurate simulation? Well, let's be practical. Since most passive components vary by 1 to 5 percent, let's say that accurate (practical) simulation requires device models with no more than 5 percent rms error over their range of operation. This is a practical compromise between the accuracy of the simulation and the amount of time we will spend making the model. For this text, a practical device model will be one which meets this 5 percent rms error criterion. In many cases, our models will be much better than this, achieving only 1 or 2 percent rms error over the entire operating range. Five percent simply defines our worst-case limit.

What does a 5 percent rms variation look like? Good question. Figure 1.2(*a*) illustrates the I-V characteristics of a MOSFET. Figure 1.2(*b*) illustrates the original curve and a 5 percent rms variation from the original. In this text, our goal is to achieve models which have 5 percent rms error or less. If you follow the guidelines outlined here, most of the models you create will have significantly less than 5 percent rms error. The curves observed in Fig. 1.2(*b*) simply represent the worst-case error which still meets our criteria for a practical device model.

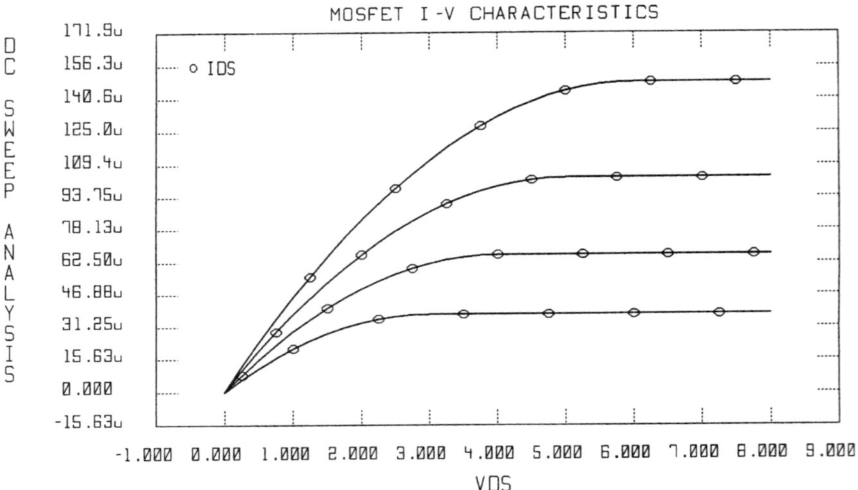

Figure 1.2(*a*) I-V characteristics of MOS transistor.

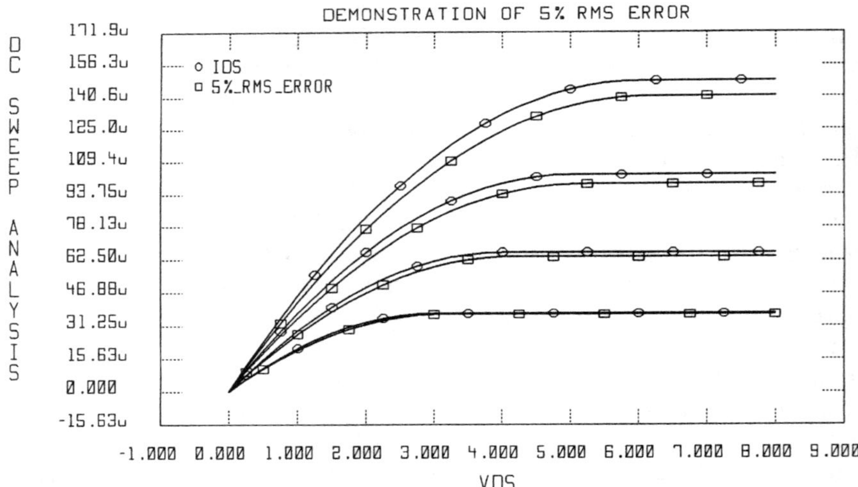

Figure 1.2(*b*) Original curves and 5 percent rms deviation.

Absolute model accuracy is an elusive goal and should probably be left to the Ph.D. candidates with time on their hands. Working engineers should try to achieve a practical level of device model accuracy. Practical device models are a compromise between absolute accuracy and reasonable accuracy, between long model development times (and costs) and reasonable model development times. The goal of this book is to teach you to develop relatively accurate device models (with less than 5 percent rms error over their operating range) in a reasonable amount of time.

For the purpose of this text, a practical model is one which exhibits less than 5 percent rms error over the entire operating range for which the model is produced, can be developed in less than one eight-hour day, and which does not exhibit any nonconvergence-causing mechanisms. You might be surprised to find out how easy it is to achieve these results.

Before proceeding with our discussion of device model accuracy, we should introduce the SPICE model hierarchy. The model hierarchy consists of three types of device models. The first of these, and the lowest of the hierarchy, is known as a *transistor-level model*. Transistor-level models are simulation models which are developed from the actual schematic of the device being simulated. For example, a transistor-level model of an operational amplifier could be constructed from the actual schematic of the op-amp, with each transistor, capacitor, and resistor entered into the simulator just as it is represented on the schematic. Figure 1.3(*a*) illustrates a transistor-level op-amp model.

A transistor-level model represents devices at the most basic simulation level possible. In many cases, the transistor-level model is the most accurate model possible for simulation. On the down side though, the transistor-level model also takes the most time to simulate.

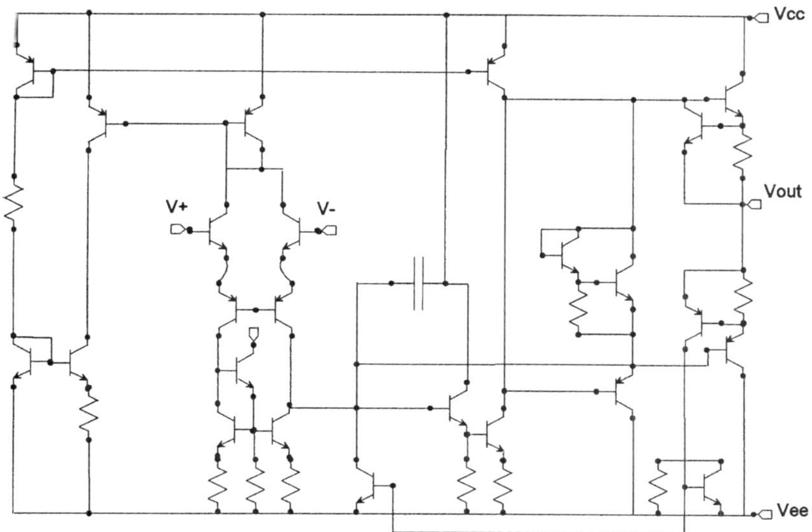

Figure 1.3(*a*) Operational amplifier transistor-level model.

The next level in the hierarchy is known as the *macromodel.* A macromodel is a collection of electrical components which form a simplified representation of the modeled circuit. Many macromodels contain dependent controlled sources to help simplify the structure of the model. Being simplified means that the macromodel is often easier to construct than the transistor-level model (unless, of course, you have a schematic), and the macromodel often simulates much faster than the transistor-level model. But these two benefits usually come at the expense of a small loss in accuracy. Figure 1.3(*b*) is a macromodel of an operational amplifier.

The highest level in the modeling hierarchy is the *behavioral macromodel.* Behavioral macromodels contain a collection of ideal electrical or mathematical components. Often, behavioral macromodels contain multipliers, dividers, and other mathematical blocks which are used to describe a function of the circuit. Being at the top of the hierarchy means the behavioral model usually simulates faster than any other type of model, but often this increased speed again comes from a loss in accuracy. Figure 1.3(*c*) illustrates a simple behavioral macromodel of an op-amp.

Each of these models is good when used in the appropriate place. Each model has distinct advantages and disadvantages. Often, good models will contain pieces of one or more levels of the hierarchy. As you will see, the models in this text will borrow from all three.

One more distinction should be mentioned here. All of the models in this text will be described as *subcircuit models.* Subcircuits allow the SPICE user to describe a model as a black box with a fixed number of inputs and outputs. Once defined, the subcircuit can easily be

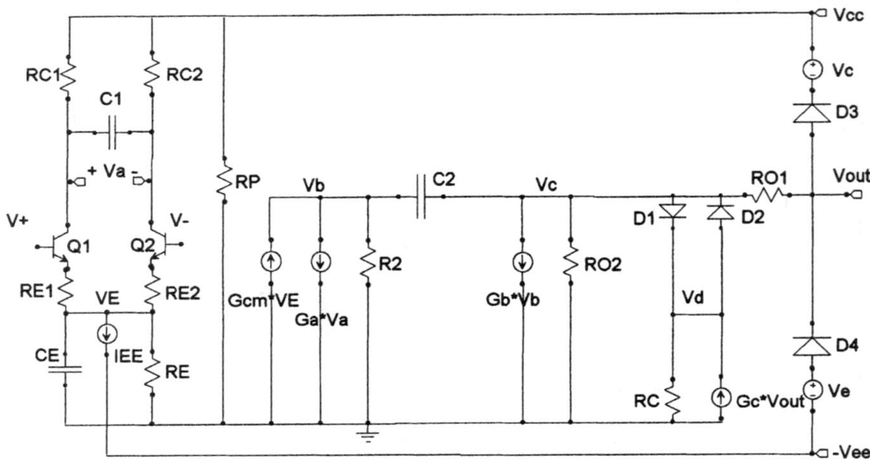

Figure 1.3(*b*) Operational amplifier macromodel.

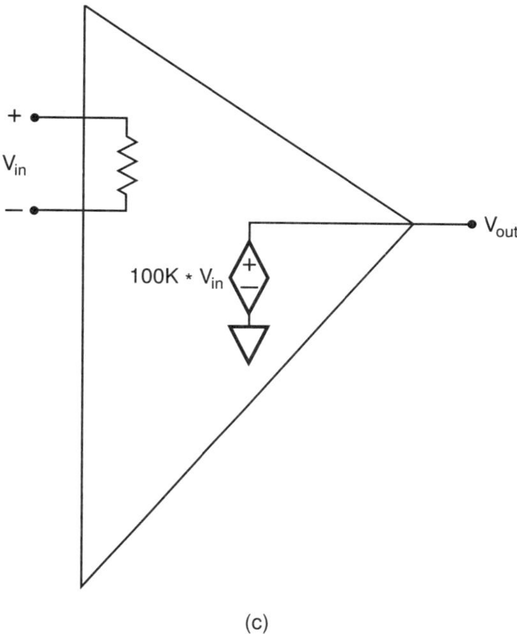

Figure 1.3(*c*) Operational amplifier behavioral model.

replicated throughout the circuit as often as required with a single, simple statement. Subcircuit models can be stored in separate files and called as needed from inside the simulator. Subcircuits offer SPICE users an easy-to-use method of building, storing, and saving device models. Readers not familiar with the concept of subcircuit modeling should review the appropriate section of their SPICE user's manuals.

The Approach to Practical Device Modeling

In this text, we will begin our model construction from standard data sheet information. You may be surprised to learn that although the typical data sheet seems to contain a wealth of information, most of the data found on a data sheet is of little value to the device modeling engineer. Even with these limitations, we will begin with the device data sheet and obtain as much modeling information as possible from them. Armed with this information, we will then use simple bench-top measurement equipment to obtain basic I-V, C-V, and Z-F characteristics. From these measured characteristics the basic model parameters and element values for the model will be obtained.

Once the final model is assembled, SPICE simulations will be performed on the newly constructed model to reproduce the I-V, C-V, and Z-F characteristics from which the parameter values were originally extracted. This will give us an opportunity to tweak the model's element values to obtain a best fit and to gauge the overall accuracy of the model.

Chapter

2

Modeling Resistors, Capacitors, and Inductors

Introduction

The need to model passive components will come as a surprise to many simulation users. SPICE contains built-in models for capacitors, inductors, and resistors. These built-in models are known as SPICE primitives. The SPICE primitives are all modeled as ideal components and, therefore, don't account for the nonideal effects caused by parasitic elements of the physical device. For many circuits, especially high-frequency circuits, an accurate simulation will require models which describe the nonideal, parasitic effects of capacitors, inductors, and resistors.

To demonstrate the shortcomings of the SPICE primitives for passive components, let's perform a simple simulation with an electrolytic capacitor. The demonstration will begin with a SPICE voltage source and a 1500-μF capacitor, as illustrated in Fig. 2.1. By measuring the

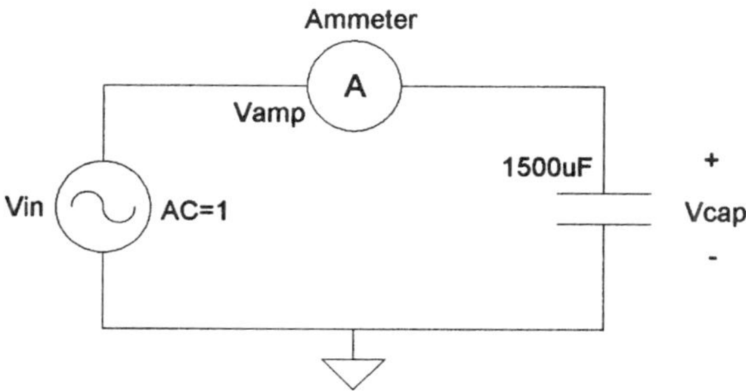

Figure 2.1 Schematic for SPICE capacitor impedance test.

current flowing into the capacitor and, at the same time, measuring the voltage across the capacitor, the impedance of the device may be determined. If we perform a frequency sweep simulation while measuring the capacitor's current and voltage, the capacitor's impedance-vs.-frequency characteristics can be examined. The SPICE netlist required to determine the impedance-vs.-frequency characteristics is shown in Fig. 2.2, while Fig. 2.3 illustrates the resulting simulation output.

In this simulation, the impedance-vs.-frequency curve follows the classical mathematical model of a capacitor, as described in Eq. 2.1, decreasing in impedance with increasing frequency. Unfortunately, real capacitors behave very differently. Figure 2.4 illustrates the *measured* impedance vs. frequency of a 1500-μF electrolytic capacitor. Notice the obvious difference between Figs. 2.3 and 2.4.

The impedance of the SPICE capacitor primitive is described by Eq. 2.1, where s is the complex frequency variable.

```
CAPACITOR TEST
VIN 1 0 AC=1
VAMP 1 2
C1 2 0 1500UF
.AC DEC 10 1 100MEG
.PRINT AC V(2) I(VAMP)
.END
```

Figure 2.2 Input file for capacitor impedance test.

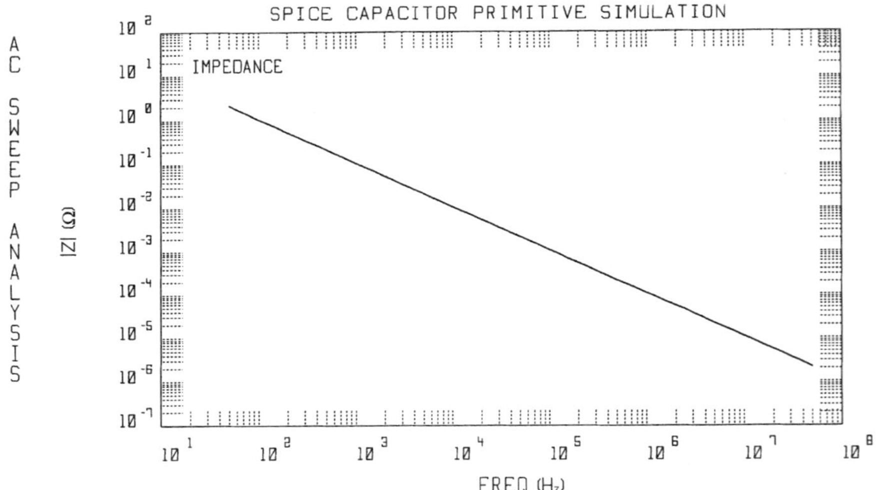

Figure 2.3 Impedance characteristics of 1500 μF SPICE primitive.

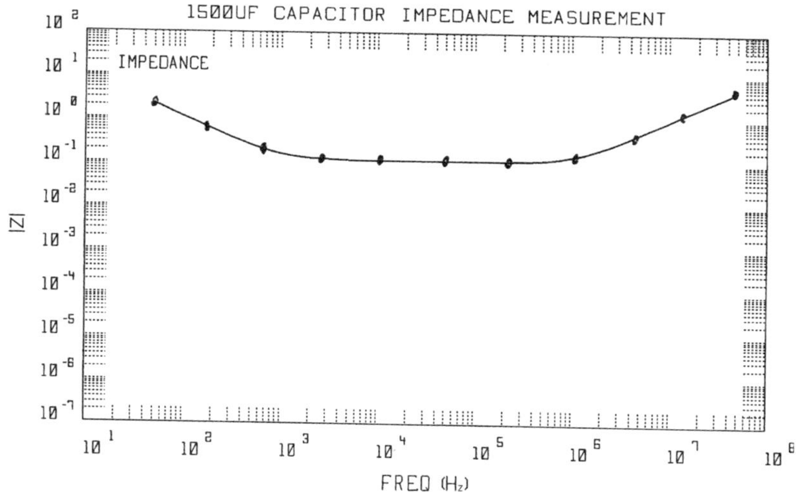

Figure 2.4 Measured impedance characteristics of 1500-μF capacitor.

$$Z = \frac{1}{sC} \tag{2.1}$$

But real capacitors *cannot* be accurately described by Eq. 2.1 over the entire frequency spectrum. Series resistance and lead inductance, nonideal parasitic byproducts of the manufacturing process, influence the impedance of the capacitor at certain frequencies. At low frequencies (up to about 1000 Hz in our example), the *measured* impedance curve matches the *ideal* impedance curve, but from 1000 Hz to about 600 kHz, the measured impedance curve reaches a minimum value and does not fall below this minimum. The minimum impedance of the capacitor is determined by its electrical series resistance (ESR). The ESR is a measure of the resistance of the contacts or leads and the resistance of the material used to make the capacitor. The value of ESR, and therefore its effect on the impedance curve, is heavily dependent on the construction and chemical composition of the capacitor. More will be said about ESR in the capacitor section of this chapter. At high frequencies (greater than 600 kHz), the impedance of the capacitor starts increasing with frequency. This effect is due to lead inductance. At very high frequencies, a capacitor actually begins looking like an inductor! Figure 2.5 illustrates how these nonideal elements modify the impedance-vs.-frequency curve.

To accurately describe these nonideal behaviors and develop an accurate capacitor model, a macromodel (subcircuit model) of the device must be constructed. The macromodel we will use for the

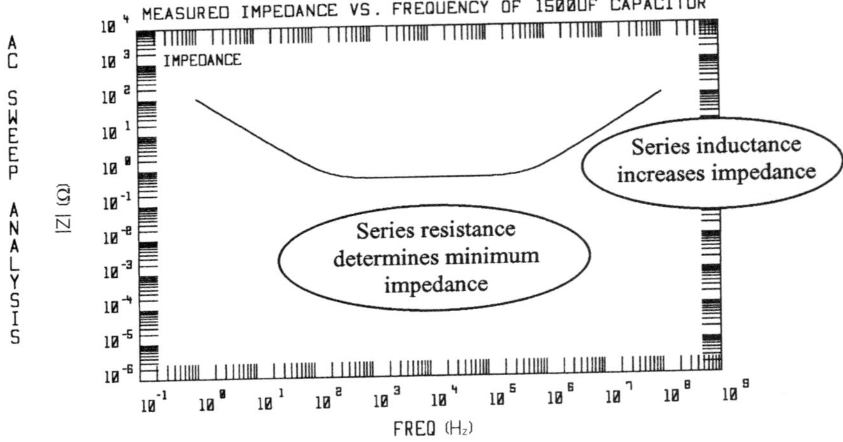

Figure 2.5 Nonideal components of impedance characteristics.

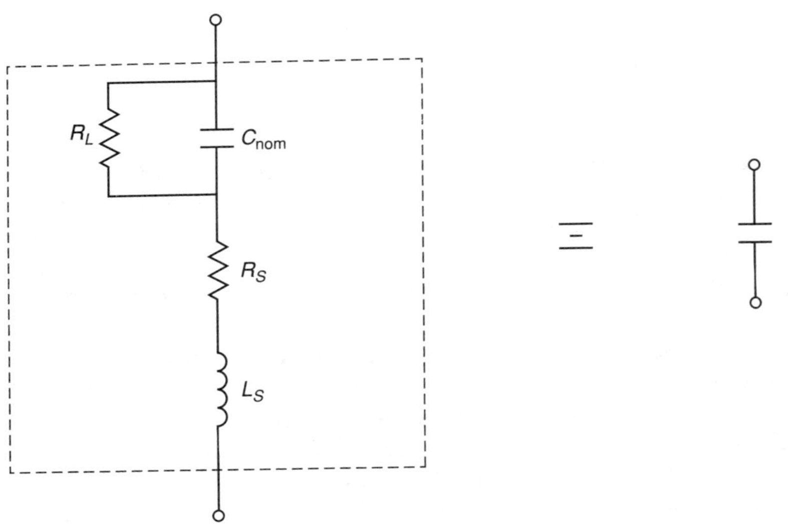

Figure 2.6 Capacitor macromodel.

capacitor is shown in Fig. 2.6; however, the detailed discussion of this macromodel will be postponed until the capacitor section of this chapter.

Replacing the capacitor primitive of our previous circuit with the capacitor macromodel yields a dramatic improvement in the simulated impedance-vs.-frequency curve. Figure 2.7 illustrates the simulated impedance-vs.-frequency curve with the capacitor macromodel.

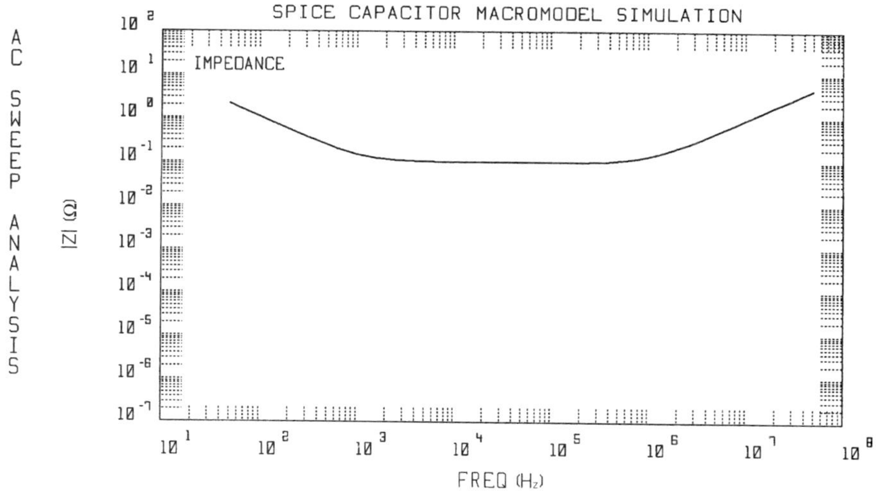

Figure 2.7 Simulation output with capacitor macromodel.

Comparing the measured capacitor impedance (Fig. 2.4) with the simulation output from the capacitor primitive (Fig. 2.3) and the capacitor macromodel (Fig. 2.7) will convince most users of the value of using macromodels for passive components.

To further illustrate the problem between the ideal SPICE capacitor primitive and the real capacitor, imagine a capacitor being used as a noise suppressor. Figure 2.8 illustrates a simple test circuit where noise spikes are added to a DC signal. The voltage spikes are being used to simulate noise in the circuit. Figure 2.9(*a*) illustrates the simulation output with a 1500-μF SPICE *capacitor primitive* used to model the capacitor, and Fig. 2.9(*b*) illustrates the simulation output when the 1500-μF *capacitor macromodel* is used to model the capacitor. In the first simulation, the ideal capacitor's high-frequency impedance effectively shunts the noise spikes to ground, leaving very little noise at the load, with no apparent effect on the voltage at the load. But the macromodel used during the second simulation exhibits

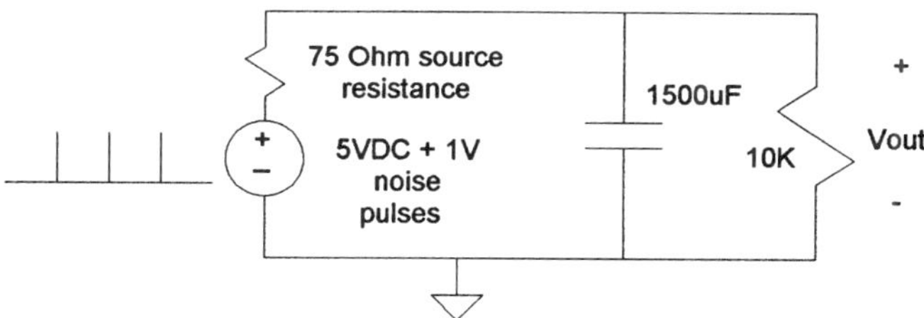

Figure 2.8 Circuit schematic for capacitor model simulation.

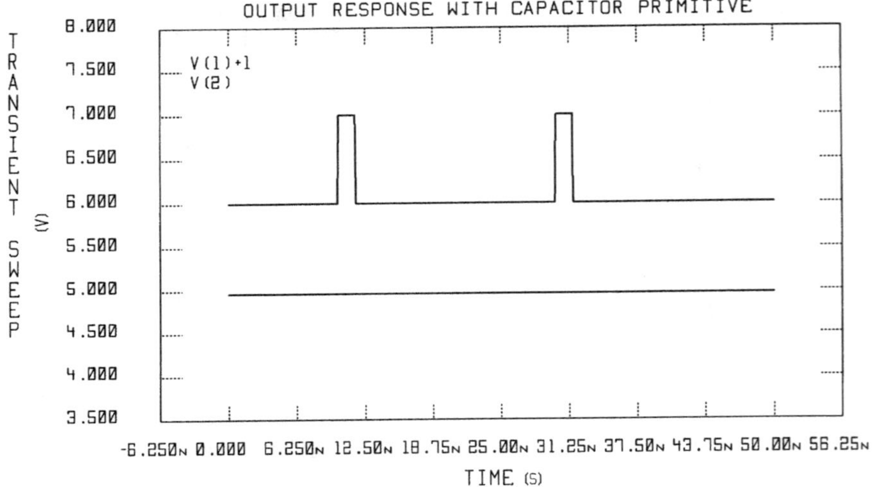

Figure 2.9(*a*) Simulation with capacitor primitive.

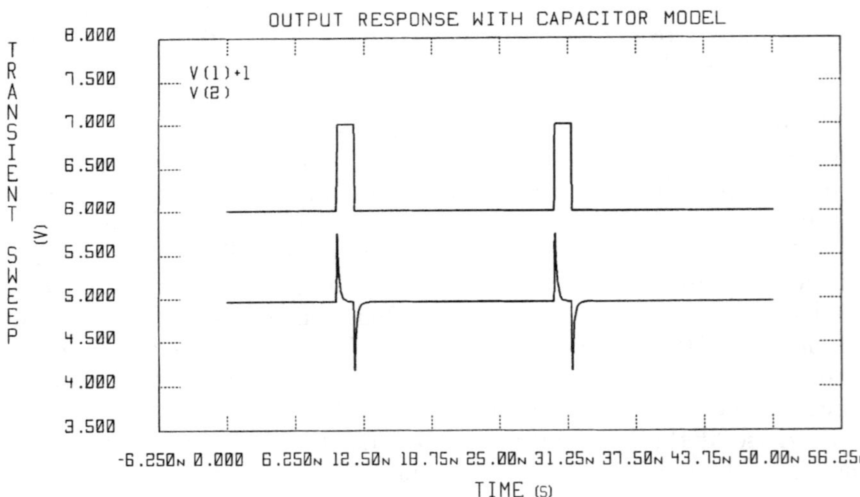

Figure 2.9(*b*) Simulation output with capacitor macromodel.

the nonideal high-frequency behavior of the real capacitor and, because of this, more of the noise reaches the load.

Figures 2.9(*a*) and (*b*) demonstrate the dramatic effect accurate macromodels can have on simulation. While most designers realize the need for accurate semiconductor models, not enough take the time to build accurate models of the passive components. If we are to achieve our goal of simulations with less than 5 percent rms error, we must utilize accurate macromodels for our passive components. Fortunately, building macromodels of resistors, capacitors, and inductors is easy and requires only a minimum investment of time and effort.

In this section, we will examine macromodels which can be used to build practical device models for capacitors, inductors, and resistors. All of the macromodels are pleasantly simple in structure, and all of the components of the macromodels can be determined with a few simple tests. We will show you how to calculate the component values for each of the macromodels. The macromodels in this section should be used whenever your circuits' components are driven into a region of operation where parasitic, nonideal effects are observed.

In this section, we will also examine the question of temperature modeling for these devices. Although SPICE offers a mechanism to model the temperature coefficient of resistance for resistors, there are no equivalent facilities for capacitors or inductors. Again, with the aid of simple macromodels, practical device models of temperature-dependent capacitors and inductors can be developed.

Macromodeling Capacitors

Capacitor technology is advancing quickly. Capacitors come in a variety of packages and with dozens of different dielectric materials. Because of this, simulating your circuit with the proper model becomes just as important as choosing a real capacitor for your circuit.

The capacitor macromodel we will use was shown previously in Fig. 2.6. The macromodel consists of a capacitor CNOM, which represents the nominal capacitance value; a series resistor RS, which represents the electrical series resistance (ESR); a series inductor LS, which represents the inductance of the device leads; and a leakage resistance RL, which represents the leakage of the capacitor. Although this model is quite simple, it can be used to accurately model both surface-mount and leaded capacitors.

To model the capacitor, the values of CNOM, RS, LS, and RL must be found. Unfortunately, many capacitor data sheets contain little or no information describing the series resistance (ESR), lead inductance, or leakage resistance. Occasionally, a manufacturer will decide that one or more of these quantities is important enough to place on the data sheet, but this occurs infrequently enough that we cannot assume the information will always be there when we need it. Because of this, we must rely on laboratory measurements to obtain accurate parameter values for the model.

To measure the element values of the capacitor macromodel, three pieces of lab equipment are required: a capacitance meter, a high-impedance resistive bridge or ohm meter, and a network analyzer or impedance meter.

The capacitance meter may be any of the common meters available today but will preferably be one of the small-signal, low-frequency (1 mV at 10 kHz) meters. The capacitance meter is used to measure the low-frequency nominal capacitance value (CNOM). If the meter uses a test frequency which is too high, the ESR or lead inductance of the capacitor will cause an erroneous reading.

The resistive bridge or ohm meter should have an input impedance as high as possible, but any with an input impedance of 10 MΩ or higher will yield satisfactory results. The resistive bridge is used to measure the leakage resistance of the capacitor. If the bridge has an input impedance which is significantly lower than the leakage resistance, obtaining an accurate measurement will be difficult.

The most expensive piece of equipment is the network analyzer. Network analyzers are designed to work over a range of frequencies. A common low frequency meter is 50 Hz to 200 MHz. The 100-kHz to 1-GHz meter is also common. The network analyzer is used to determine the impedance-vs.-frequency characteristics of the capacitor. Once determined, the ESR (RS) and lead inductance (LS) macromodel values will be calculated from the impedance characteristics. Although the capacitor's impedance-vs.-frequency characteristics will vary depending on the construction of the capacitor, network analyzers at the lower end of the frequency spectrum are usually desirable for this test.

If a network analyzer is not available, the impedance-vs.-frequency characteristics can be determined with an impedance meter. Impedance meters usually have a very high input impedance and can measure the impedance of a device over a range of frequencies.

Begin the model extraction procedure by measuring the low-frequency capacitance value with the capacitance meter. The test setup is shown in Fig. 2.10. Record this value as CNOM, which represents

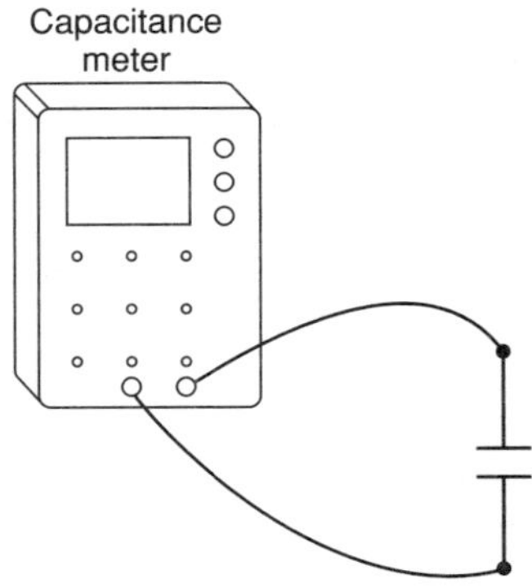

Figure 2.10 Measuring the nominal capacitance for capacitor macromodel.

the nominal capacitance value of the macromodel. Set CNOM of the capacitance model to this value.

If a low-frequency capacitance meter cannot be found, the value of CNOM can be determined from the impedance-vs.-frequency plot we will generate with the network analyzer. More will be said about this later.

After the nominal capacitance value has been determined, the ESR (RS) and lead inductance (LS) values should be determined. Both of these quantities are determined from impedance-vs.-frequency plot, and the plot is obtained with the network analyzer. The test setup to determine the impedance-vs.-frequency curve is shown in Fig. 2.11.

When measuring the impedance-vs.-frequency curves, the capacitor is placed across the source and measure leads of the analyzer. When measuring the impedance of the capacitor, the input impedance of the network analyzer must be considered. The analyzer we will use has a 50-Ω input impedance. It can be seen from Fig. 2.12(*a*) that in the case of our 1500-μF capacitor, the capacitor's impedance is much less than 50 Ω from 50 Hz to 50 MHz and, therefore, the low input impedance of the network analyzer does not present a measurement problem.

Figure 2.12(*a*) illustrates the measured impedance-vs.-frequency plot of our 1500-μF capacitor. It is important to note that capacitors which exhibit extremely large values of electrical series resistance usually generate an impedance-vs.-frequency characteristic similar to the one shown in Fig. 2.12(*a*), but capacitors which have a low ESR value will have an impedance-vs.-frequency curve similar to Fig. 2.12(*b*). The capacitor's ESR value will have a pronounced effect on the shape of the impedance-vs.-frequency curve. This effect is illustrated in Figs. 2.13(*a*), (*b*), and (*c*).

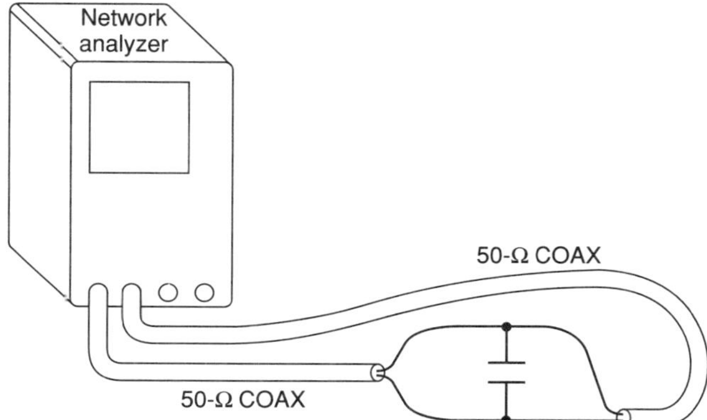

Figure 2.11 Using a network analyzer to determine impedance characteristics of a capacitor.

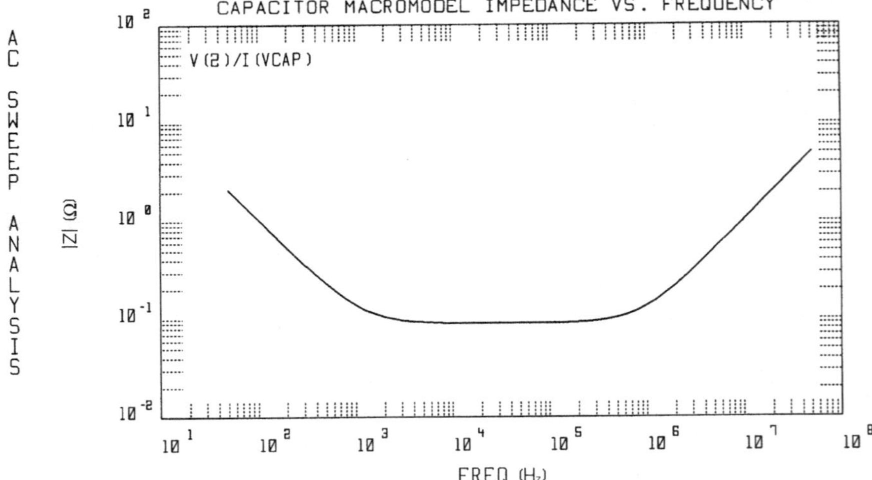

Figure 2.12(*a*) Impedance characteristics of high-ESR capacitor.

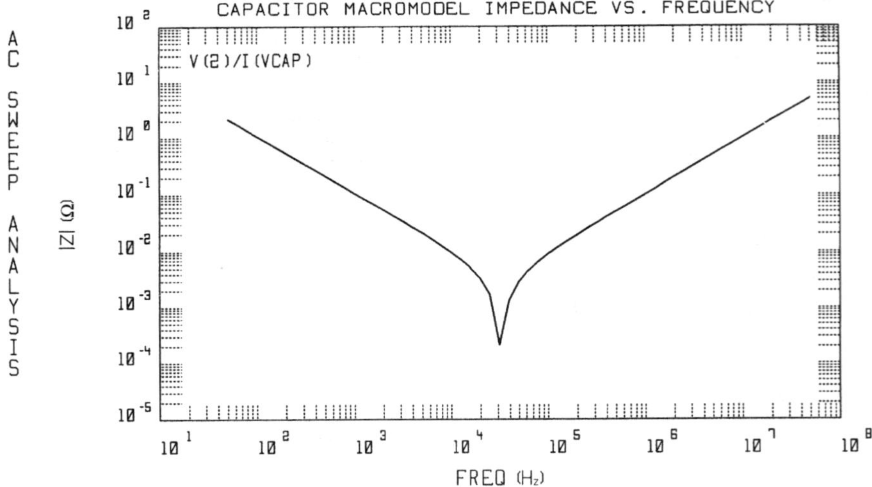

Figure 2.12(*b*) Impedance characteristics of low-ESR capacitor.

The ESR of the capacitor macromodel (RS) is measured from the minimum impedance of the impedance-vs.-frequency curve. Determine the minimum impedance as shown in Fig. 2.14, and set RS to this value.

If you have not determined the nominal capacitance value (CNOM) because you do not have access to a low-frequency capacitance meter, the impedance-vs.-frequency curves offer another way to determine this value. Equation 2.2 illustrates the relationship between the mag-

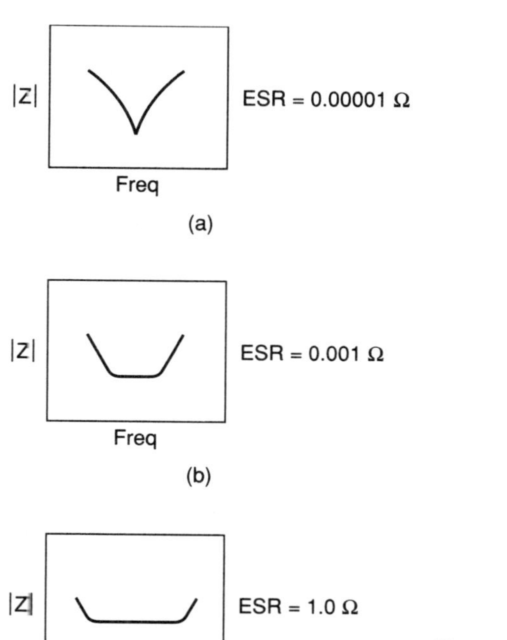

Figure 2.13 Impedance characteristics of low, medium, and high ESR capacitors.

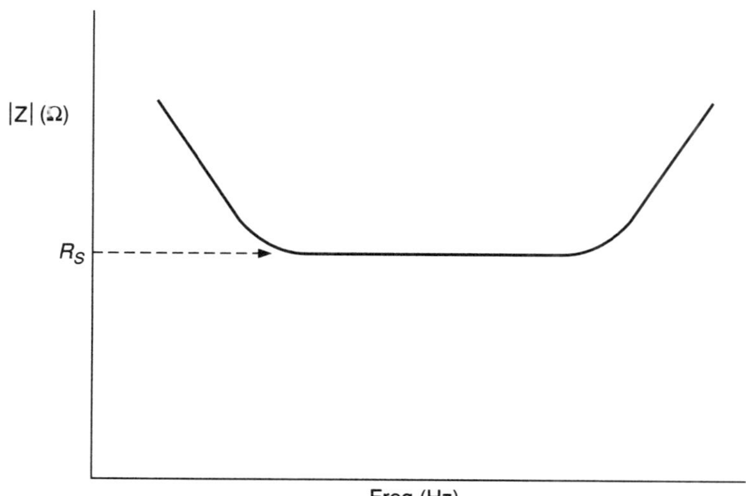

Figure 2.14 Calculating the minimum impedance (RS).

nitude of the impedance, the nominal capacitance, and the frequency of operation for the ideal capacitor. On the impedance-vs.-frequency plot, the capacitor behaves as described by Eq. 2.2 until approximately 1 kHz, where the capacitor's ESR value becomes the dominant impedance component. Figure 2.15(*a*) illustrates the regions of ideal and nonideal capacitance behavior. By measuring the impedance value at a given frequency within the ideal region, the nominal capacitance value can be determined from Eq. 2.3. This measurement is illustrated in Fig. 2.15(*b*).

$$|z| = \frac{1}{\omega C} = \frac{1}{2\pi f C} \tag{2.2}$$

$$\text{CNOM} = \frac{1}{2\pi f |z|} \tag{2.3}$$

The series inductance component of the model (LS) is determined from the self-resonating frequency F_{SRF} of the capacitor. The self-resonating frequency is the point at which the capacitor's lead inductance starts to impact the impedance-vs.-frequency curve. The self-resonating frequency is determined from the impedance-vs.-frequency curve in one of two ways depending on the shape of the curve. For capacitors with a very low ESR value, the self-resonating frequency is determined from the intersection of the two straight-line segments of the curve as illustrated in Fig. 2.16(*a*). For capacitors with larger values of ESR, the self-resonating frequency can be determined by

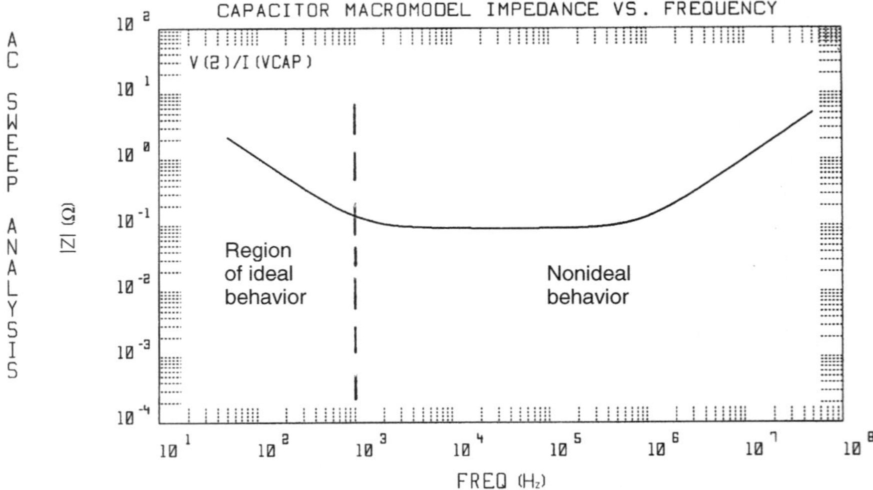

Figure 2.15(*a*) Ideal and nonideal regions of behavior for real capacitor.

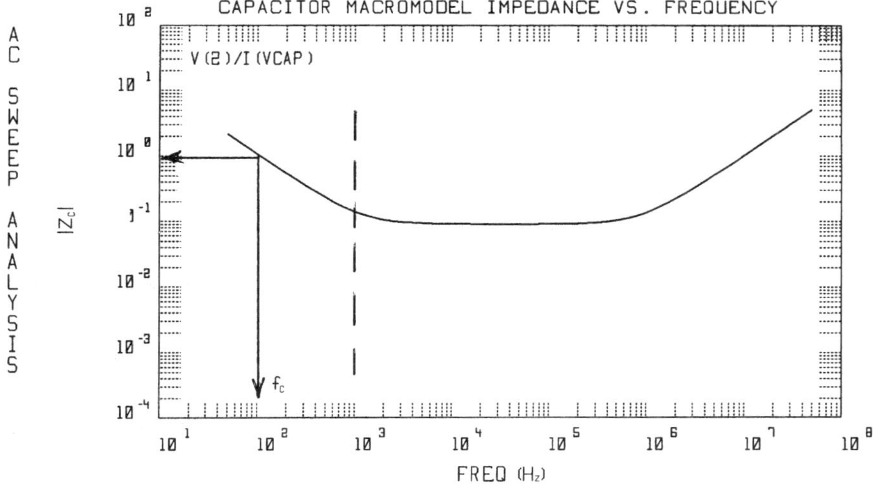

Figure 2.15(*b*) Determining the nominal capacitance (CNOM) for the capacitor macromodel.

extending the linear portion of the curves and locating their intersection, as illustrated in Fig. 2.16(*b*).

The self-resonating frequency is the frequency at which the reactance of the nominal capacitance is equal (and opposite in sign) to the reactance of the series inductance. At this frequency, the LC components of the device are resonating. Once the self-resonating frequency

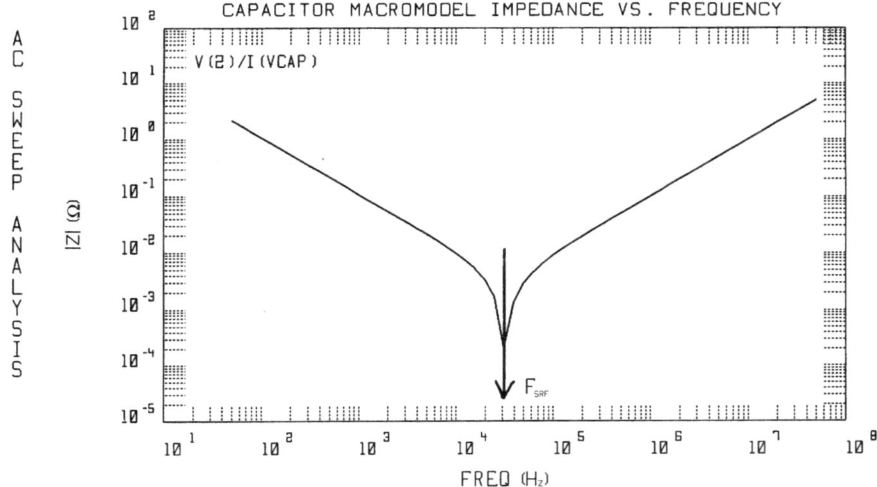

Figure 2.16(*a*) Determining the self-resonating frequency of a low-ESR capacitor.

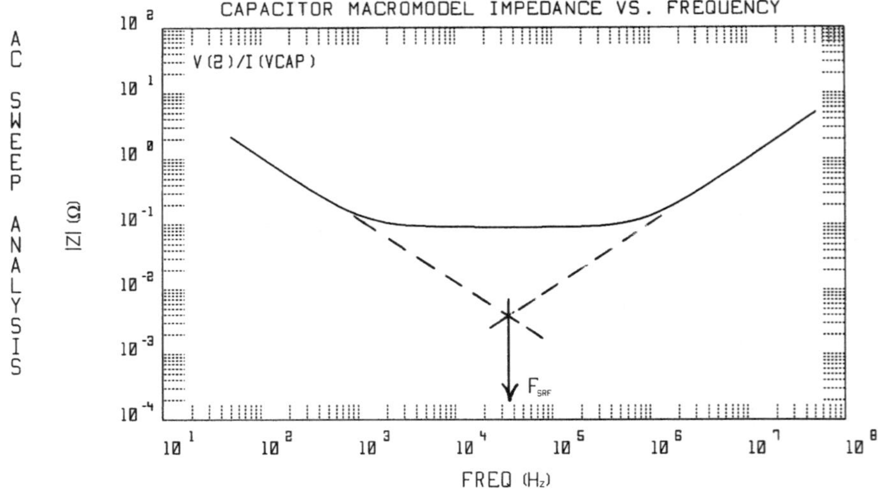

Figure 2.16(*b*) Determining the self-resonating frequency of a high-ESR capacitor.

has been found, the series inductance (LS) can be determined from the classic resonance relationship for an LC circuit. Equation 2.4 describes the resonant frequency and radian resonant frequency of an LC circuit, and, from this, Eq. 2.5 can be derived and used to determine the series inductance term (LS). Determine the self-resonating frequency from the impedance-vs.-frequency plot, use Eq. 2.5 to solve for LS, then set LS of the capacitance model to this value.

$$\omega = 2\pi f_{\text{SFR}} = \frac{1}{\sqrt{\text{LC}}} \tag{2.4}$$

$$\text{LS} = \frac{1}{4\pi^2 f_{\text{SRF}}{}^2 C} \tag{2.5}$$

The leakage resistance (RL) is the final element we need to develop the capacitance macromodel. But before we begin the discussion of how to extract the leakage resistance, it should be noted that, for many devices, the leakage resistance will have very little influence on the behavior of the macromodel. Many capacitors can be accurately modeled either by setting RL to large value, say 10 MΩ, or by removing the RL term completely. However, for charge-storage circuits, switch capacitor filters, and other charge-sensitive circuits, the RL term is vital in producing an accurate simulation.

Measuring the leakage resistance requires a high-impedance resistive bridge or ohm meter. A meter with an input impedance of 10 MΩ

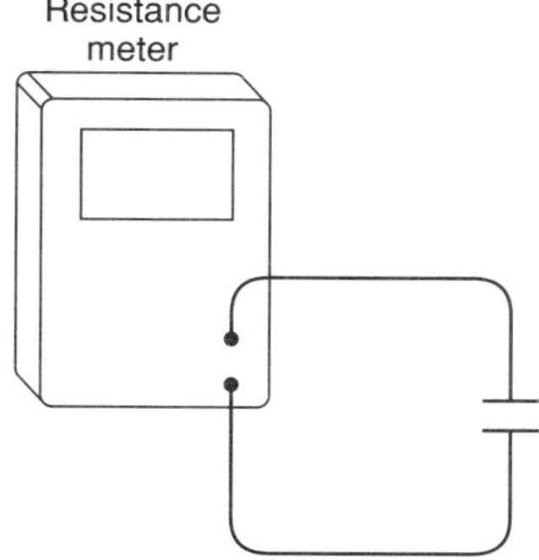

Figure 2.17 Measuring the leakage resistance of a capacitor.

or greater is desirable for this measurement. To measure the leakage resistance, simply place the capacitor across the bridge or meter, as shown in Fig. 2.17, and measure the leakage resistance. Set RL of the capacitance macromodel to the measured value.

Once the measurements have been completed, the model should be verified through simulation. In the `\RCGV34\CH2` subdirectory of the disk which accompanies this text, there is a capacitor macromodel netlist and an RSPICE circuit netlist to simulate the impedance-vs.-frequency characteristics of the capacitor model. The capacitor macromodel netlist is stored in the file named CAP.MOD. Also included in this subdirectory is a batch file which automates the impedance-vs.-frequency simulation/graphing procedure. The batch file is named CAP.BAT.

To test the capacitor macromodel, change into the `\RCGV34\CH2` subdirectory and edit the capacitor macromodel netlist CAP.MOD. Set the CNOM, RS, LS, and RL terms to appropriate values. Save the CAP.MOD file, and simulate the circuit with the CAP.BAT batch file by typing:

```
CAP.BAT
```

Your results should be similar to those shown in Fig. 2.18.

The batch file will simulate the CAP.MOD subcircuit and plot the impedance-vs.-frequency characteristics from 50 Hz to 50 MHz on the screen. When the impedance curves appear on the screen, the batch file returns control to the user, whereby the user may obtain a hardcopy of the curve or exit the program.

Together, the CAP.MOD macromodel and the CAP.BAT batch file allow the user to tweak the CNOM, RS, LS, and RL terms until a precise fit is obtained between the measured data and the simulated data.

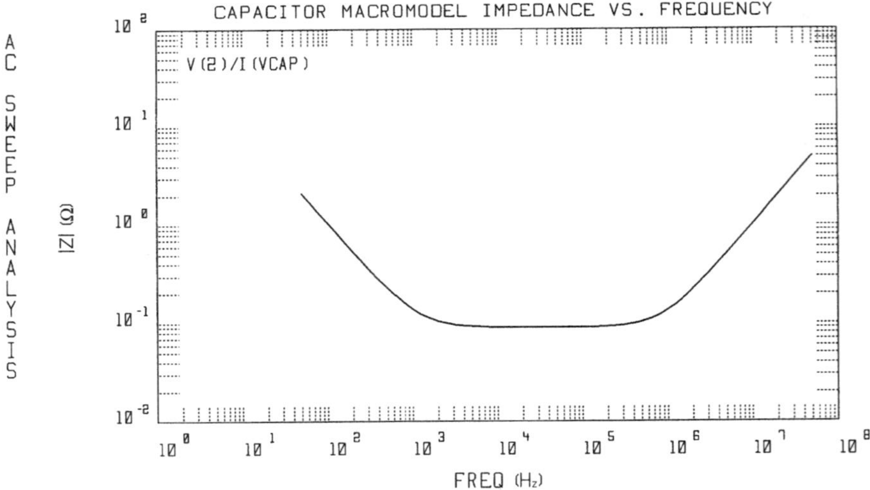

Figure 2.18 Typical capacitor impedance characteristics.

Capacitor Macromodel Demonstration

To demonstrate the capacitor macromodeling technique, a 68-μF tantalum electrolytic capacitor was selected. Because our network analyzer was on the *blink* (technical term), the impedance-vs.-frequency characteristics were determined with a simple impedance meter. Table 2.1 illustrates the measured values.

The data in Table 2.1 was entered into a text file and plotted with the RGRAPH postprocessor which accompanies this text. Figure 2.19(*a*) illustrates the measured impedance-vs.-frequency characteristics of the capacitor. The value of impedance at 1000 Hz was used to determine the nominal capacitance value. Using 1000 Hz, 2.30 Ω, and Eq. 2.3 yields a nominal capacitance of 69.2 μF.

Using Fig. 2.19(*a*) and the technique shown in Fig. 2.16(*b*), the self-resonating frequency F_{SRF} was determined to be 108 kHz. With F_{SRF}

TABLE 2.1 Measured Impedance-vs.-Frequency Characteristics

Frequency	Impedance magnitude, Ω
10 Hz	223.0
100 Hz	22.0
1000 Hz	2.30
10 kHz	.245
100 kHz	.092
1 MHz	.249
10 MHz	2.13

set at 108 kHz, Eq. 2.5 prescribes a series inductance of 32 μF. The minimum impedance shown in Fig. 2.19 is .092 Ω, and RS of the model was set to this value. These values were entered into the capacitor macromodel template, simulated, and compared with the measured characteristics. The measured and simulated characteristics are shown in Fig. 2.19(*b*). Look closely—you may have a hard time discerning the two different curves. The RGRAPH error function

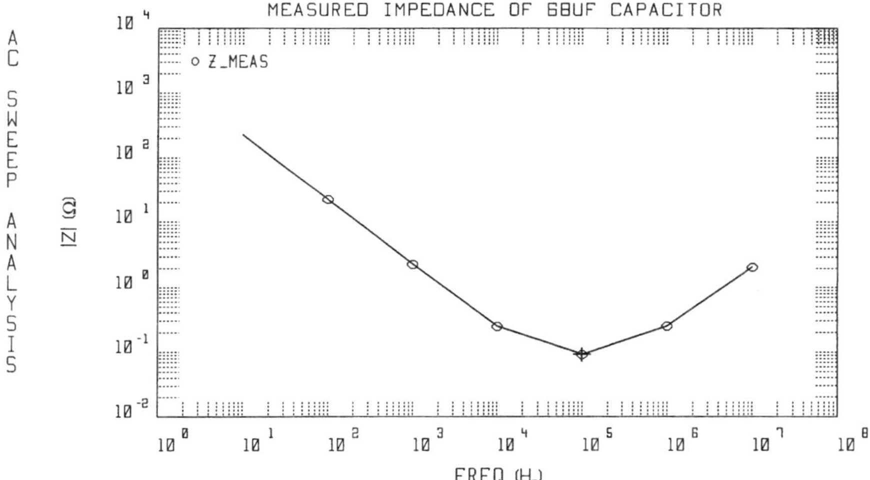

Figure 2.19(*a*) Impedance characteristics of 68-μF capacitor.

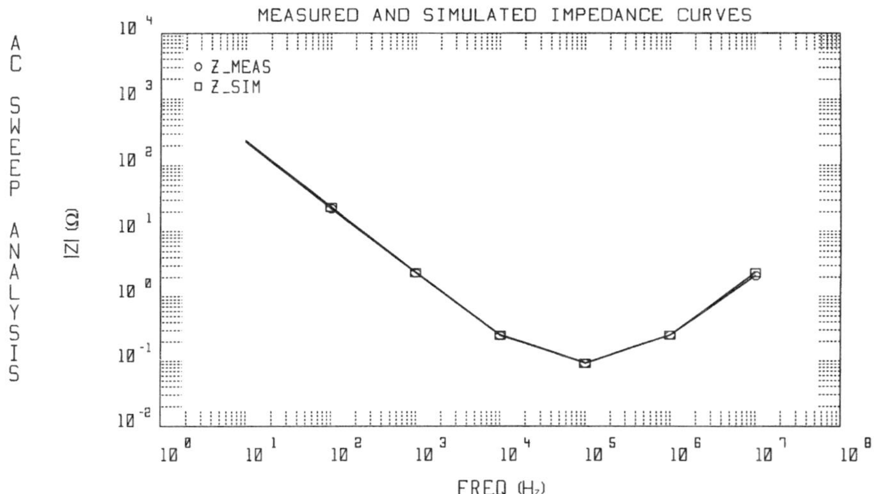

Figure 2.19(*b*) Measured and simulated impedance curves.

ERR (x,y) can be used to evaluate the difference between two curves and the rms error of a set of data. The ERR (x,y) function generates a curve on the screen which represents the difference between data points and a report file which contains the rms error of the data set. Figure 2.19(c) illustrates the measured and simulated curves and the difference between data points. Figure 2.20 illustrates the report file which displays the rms error. As can be seen from Fig. 2.20, with only seven data points, we have created a model which is almost an exact representation of the original device. Total rms error between the simulated model and the actual device is 4.25 percent.

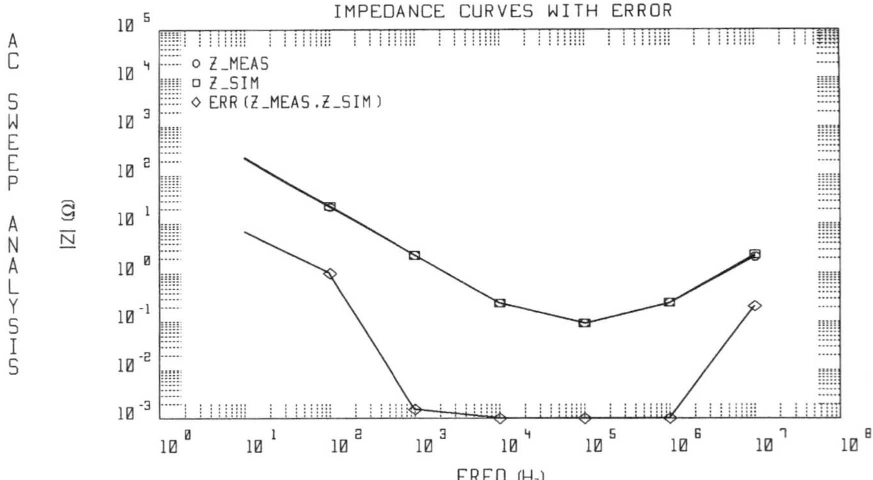

Figure 2.19(*c*) Impedance curves with error.

Error report

'Z_MEAS'	'Z_SIM'	Percent Error
22.3000E+01	22.9993E+01	3.1%
22.0000E+00	22.9994E+00	4.5%
23.0000E-01	23.0153E-01	0.1%
24.5000E-02	24.5542E-02	0.2%
92.0000E-03	92.0008E-03	0.0%
24.9000E-02	24.9050E-02	0.0%
21.3000E-01	23.3893E-01	9.8%

RMS Error = 4.25%

Figure 2.20 Error report for capacitor macromodel.

The data needed to construct a capacitor model is easy to obtain, and the steps needed to transform the measured data into a set of macromodel elements is almost trivial. Yet with minimal investment in time and effort, a significantly better simulation can be obtained. Accurate simulation does not happen by accident. You must pay attention to the details of your models. But paying attention to details does not mean you must spend an inordinate amount of time developing models. If you follow the guidelines outlined in this text, you will find that, for a moderate amount of effort, you can achieve a highly accurate, highly detailed, highly reliable simulation result—if you follow the guidelines.

Capacitive Temperature Effects

Since SPICE does not offer a mechanism to model temperature dependence of the capacitance primitive, the desire to model this behavior can only be met with a temperature-dependent macromodel. The temperature-dependent macromodel uses the same basic subcircuit schematic as shown in Fig. 2.6, but with a temperature-dependent controller added to modify the nominal capacitance and series resistance of the macromodel.

To build the temperature-dependent macromodel, the capacitor primitive (CNOM) and the resistor primitive (RS) of the model must change as a function of temperature. To achieve this, CNOM is replaced with a voltage-controlled capacitor macromodel, as shown in Fig. 2.21. The resistor's temperature coefficients will be used to describe the resistance change with temperature.

The voltage-controlled capacitor is a macromodel which will be nested inside of the capacitor macromodel. (Although the voltage-controlled capacitor model will be used here without further development, a complete derivation of this model can be found in App. D.) The voltage-controlled capacitor has the capacitance-voltage relationship expressed in Eq. 2.6.

$$C(_{\mathrm{T}}) = \mathrm{CNOM} * V_{\mathrm{CONTROL}} \tag{2.6}$$

Equation 2.7 is used to describe our temperature-dependent capacitor macromodel.

$$C(_{\mathrm{T}}) = \mathrm{CNOM} * [1 + \mathrm{TC}_1 (T - T_{\mathrm{NOM}}) + \mathrm{TC}_2 (T - T_{\mathrm{NOM}})^2] \tag{2.7}$$

By inspecting Eqs. 2.6 and 2.7, it should be apparent that, to arrive at the desired capacitance-temperature relationship, we will need an element or model which has the voltage-temperature relationship described in Eq. 2.8

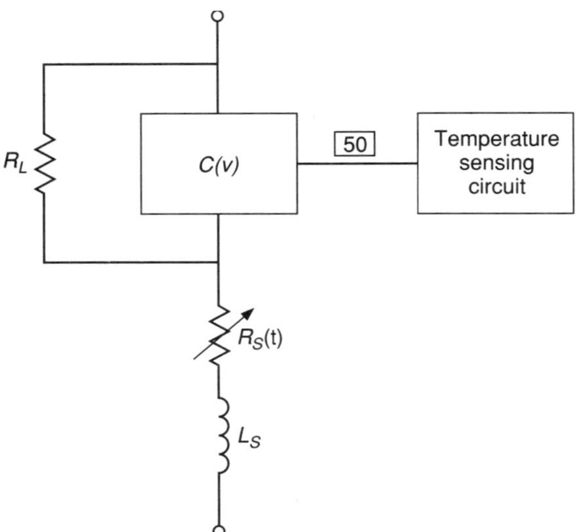

Figure 2.21 Temperature-dependent capacitor macromodel.

$$V_{\text{OUT}} = 1 + \text{TC}_1\,(T - T_{\text{NOM}}) + \text{TC}_2\,(T - T_{\text{NOM}})^2 \tag{2.8}$$

or expressed in terms of the control voltage, Eq. 2.9.

$$V_{\text{CONTROL}} = 1 + \text{TC}_1\,(T - T_{\text{NOM}}) + \text{TC}_2\,(T - T_{\text{NOM}})^2 \tag{2.9}$$

Substituting Eq. 2.9 into Eq. 2.6 yields the desired relationship between the capacitance and the temperature of the circuit. But we appear to be at an impasse, for SPICE does not have an element with the voltage-temperature relationship described in Eq. 2.9. To our rescue, comes the white knight known as *analog behavioral modeling* (ABM).

ABM allows us to manipulate voltages, currents, and even temperatures with mathematical relationships. ABM is best described as modeling with functional blocks. ABM blocks include adders, subtracters, multipliers, integrators, and so on. All of these blocks are build from standard SPICE components (App. D will cover building analog behavioral models), but using ABM blocks requires no knowledge about how the blocks were constructed. So for now, let us assume that someone has assembled a library of behavioral adders, subtracters, etc., and saved these blocks in a library for our use. (The `RCGV34\LIB` subdirectory of the disk which accompanies this text has a library of behavioral building blocks.)

With the knowledge of the models in our ABM library, a block diagram of the temperature-dependent capacitor model can be constructed.

The schematic of the temperature-sensing macromodel (TSM), is shown in Fig. 2.22. The model generates an output voltage (at node 50) prescribed by Eq. 2.8. In this circuit, a current source and two resistors are used to help convert the temperature of the circuit into a voltage (at node 100). When the temperature of the circuit is 27 C, node 100 is 27 V; when the temperature is −40 C, node 100 is −40 V. The voltage at node 100 is always equal to the temperature of the circuit in degrees C.

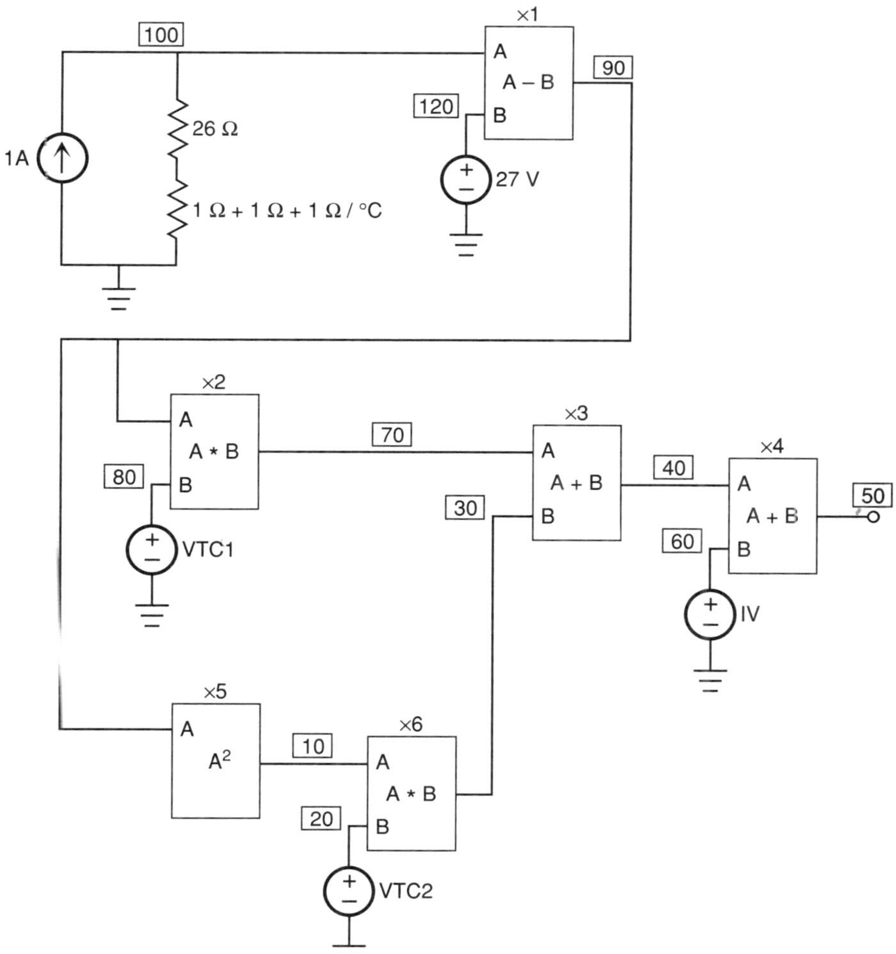

Figure 2.22 Diagram of temperature-sensing macromodel.

The voltage at node 100 is passed through several behavioral blocks. The subtracter block (X1) subtracts a constant 27 V from the voltage at node 100. The resulting voltage at node 90 is equal to the $V(100) - 27$ V or temperature -27 C (where 27 C is the nominal temperature in SPICE).

The voltage at node 90 is multiplied (X2) by the voltage source VTC1. VTC1 is a DC source whose value is set to the linear temperature coefficient of the cap. The voltage at node 70 can be described as [$V(100) - 27$ volts] times the VTC1 voltage or VTC1 *(temperature − 27 C). The voltage at node 90 is also passed into a squaring block (X5), and the output of this block, at node 10 can be described as $[V(100) - 27\ \mathrm{C}]^2$. This voltage is then multiplied by the VTC2 voltage source, and the resulting voltage at node 30 can be described as VTC2 $*(\text{temperature} - 27\ \mathrm{C})^2$. VTC2 is a DC source whose value is set to the square-law temperature coefficient of the cap. A two-stage adder adds the voltage at node 30 and node 70 to a constant 1-V source. The voltage output at node 50 is described by Eq. 2.10.

$$V(50) = [1 + TC1\,(V(100) - 27\ \text{volts}) + TC2\,(V(100) - 27\ \text{volts})^2] \quad (2.10)$$

In the `RCGV34\LIB` subdirectory of the disk which accompanies this text is a file named TSM.MOD. TSM stands for temperature-sensing model. The circuit just described is the TSM model. This model will be used several times throughout this text to help build temperature-dependent models.

The temperature coefficients (TC1, TC2) of the macromodel are set with the DC voltage of the VTC1 and VTC2 voltage sources. (A program has been written for you to help you determine the VTC1 and VTC2 temperature coefficients of your model. The program is called TC_CALC and is stored in the `UTIL` subdirectory. Refer to the section of this chapter labeled "Calculating Temperature Coefficients for Passive Components" for instructions on how to use TC_CALC.) Set the DC voltage of the source equal to the desired temperature coefficient. Use Table 2.2 to determine the proper voltage setting.

TABLE 2.2 Voltage Setting for Various Temperature Coefficients

TC (ppm)	DC voltage
10	.00001 (10 μV)
100	.0001 (100 μV)
1000	.001 (1000 μV)
−1000	−.001 (−1000 μV)
−100	−.0001 (−100 μV)
−10	−.00001 (−10 μV)

Coupling the output of the TSM with Eq. 2.6 and a 1500-μF voltage-controlled capacitor gives us the capacitance-temperature relationship shown in Eq. 2.11.

$$C(T) = 1500\ \mu\text{F}\ [1 + \text{TC1}\ (T - 27\ \text{C}) + \text{TC2}\ (T - 27\ \text{C})^2] \quad (2.11)$$

Equation 2.11 describes the temperature dependence of CNOM for the temperature-dependent capacitor macromodel. But in addition to temperature dependence of the nominal capacitance, the electrical series resistance (RS) may also change with temperature. This change can be accounted for with the standard SPICE temperature coefficients (Rxxx n1 n2 R_value TC = TC1, TC2) for resistors.

The temperature-dependent capacitor macromodel is stored in the `\RCGV34\CH2` subdirectory of the disk which was included with this text. The model is stored in the file named TCAP.MOD. To demonstrate the capacitance change over temperature, a batch procedure has been created for you. The batch file simulates the impedance-vs.-frequency characteristics at −40 C, 0 C, 40 C, 80 C, and 120 C, and superimposes all of the curves on one plot. The batch file is named TCAP.BAT. To execute the batch file, change to the `\RCGV34\CH2` subdirectory and type:

```
TCAP.BAT
```

The result of the batch file simulation is shown in Fig. 2.23.

To many, it may seem that macromodeling capacitors is overkill for simulation. However, for many circuits (especially high-power cir-

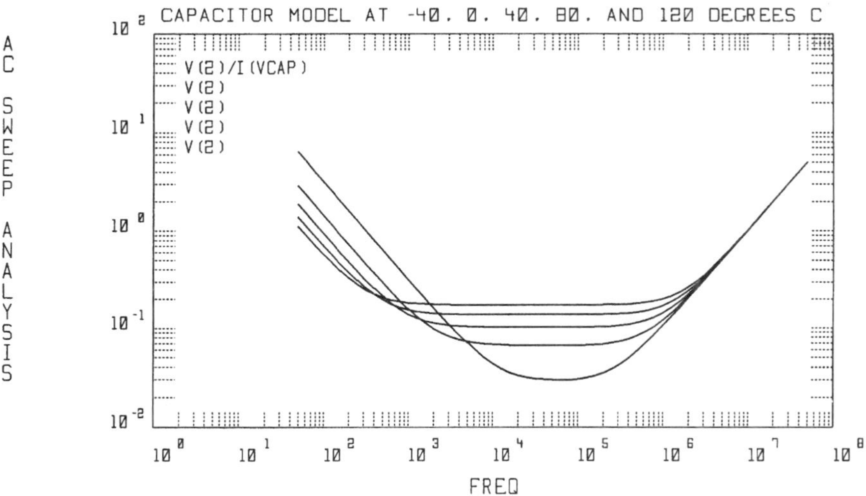

Figure 2.23 Capacitor impedance characteristics at five different temperatures.

cuits, digital circuits, noise suppression circuits, high-voltage, and high-current circuits), an accurate simulation—one within the 5 percent rms error tolerance—will require accurate capacitor macromodels. Circuits which are simulated over a range of temperatures will not accurately reflect the true behavior of the modeled circuit without temperature-dependent capacitor macromodels. Although an investment in time is required to develop these models, the return is immeasurable in terms of improving the accuracy of your simulation.

Macromodeling Inductors

Like capacitors, inductors come in a wide variety of shapes and sizes and with numerous core materials. Like the capacitor, the inductor changes behavior with frequency and temperature. For many circuits to achieve a goal of 5 percent rms error or less in our simulations, inductor macromodels must be used in place of the standard SPICE inductor primitive.

Figure 2.24(*a*) and (*b*) illustrate the differences between the behavior of the SPICE inductor primitive and a real inductor. The impedance of the ideal SPICE primitive is described in Eq. 2.12. But, because of the coil resistance and winding capacitance of the physical inductor, the accuracy of Eq. 2.12 becomes suspect at very low and very high frequencies. At low frequencies, the coil resistance, which is not included in Eq. 2.12, determines the low-frequency impedance of the coil. At very high frequencies, the winding capacitance (again not accounted for in Eq. 2.12) forces the impedance characteristics to

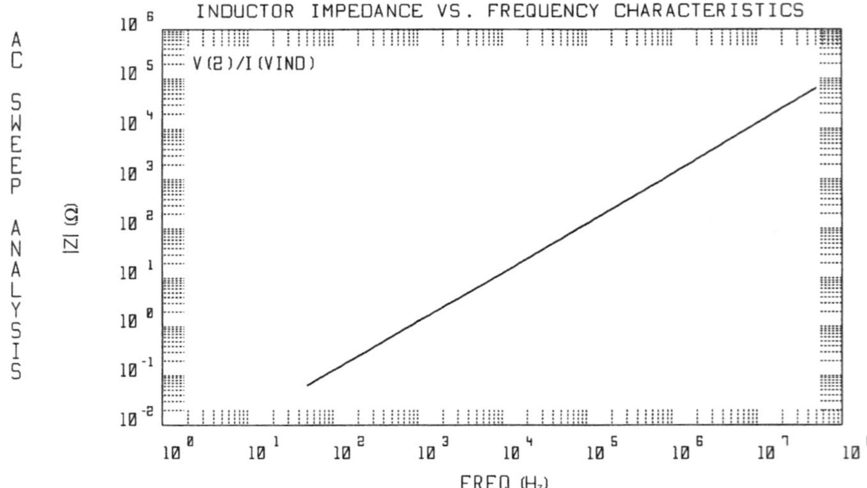

Figure 2.24(*a*) Impedance characteristics of *ideal* inductor.

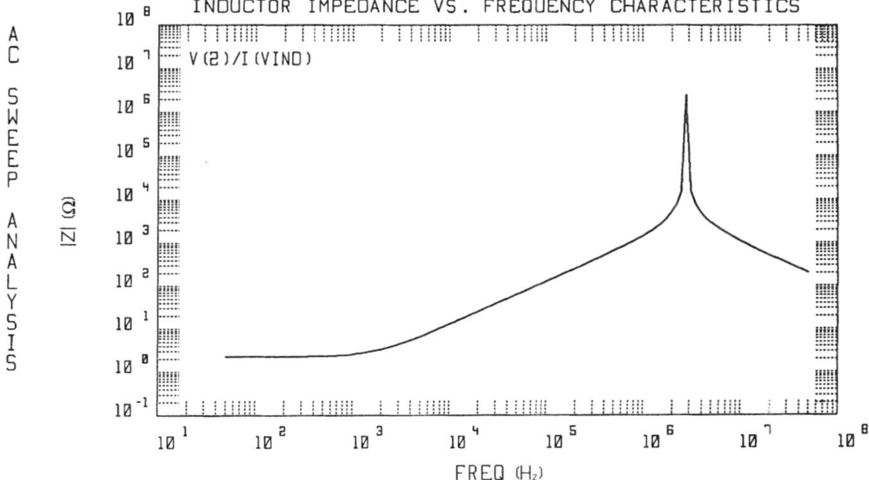

Figure 2.24(*b*) Impedance characteristics of *real* inductor.

deviate from Eq. 2.12. To account for all of these nonideal behaviors of the device, an inductor macromodel will be developed.

$$Z = SL \tag{2.12}$$

To those of you who have gone through the capacitor section, the inductor macromodel we will use should look quite familiar to you. For those of you who couldn't wait and jumped straight to this section, the inductor macromodel is shown in Fig. 2.25.

The inductor macromodel consists of three SPICE primitives, an inductor LNOM, which represents the nominal inductance; a series resistor RS, which represents the coil resistance; and a parallel capacitance CP, which represents the winding capacitance. Together, these

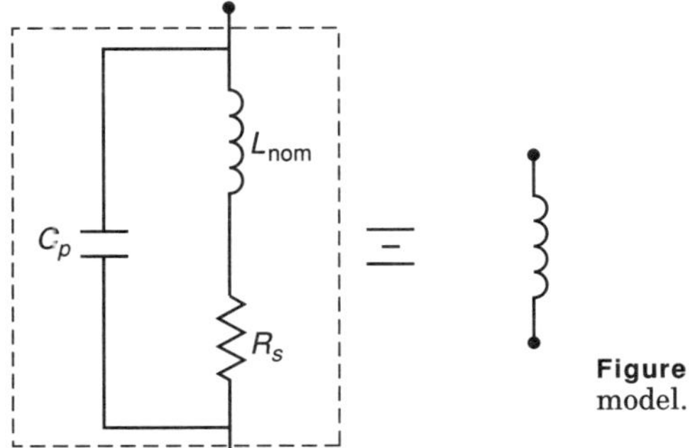

Figure 2.25 Inductor macromodel.

three elements will model the frequency-dependent behavior of a physical inductor.

In keeping with our model development philosophy, to determine the values of LNOM, RS, and CP, we should inspect the inductor's data sheet. After inspecting several data sheets, we found: (1) all of them contained information about the nominal inductance LNOM, (2) only about half of the data sheets specified any information about the coil resistance RS, and of those that did, most only specified a *maximum coil resistance,* and (3) none of the data sheets contained any information about the winding capacitance.

As in the case of the capacitor, the most reliable method of determining the values needed for the inductor macromodel involves making laboratory measurements on representative devices. The measurement equipment used for these measurements includes an impedance meter and a DC ohm meter.

The impedance meter can be any of those currently available, but should offer a mechanism to change the impedance measurement frequency. For the inductor macromodel, the impedance meter will be used to determine the impedance-vs.-frequency characteristics of the device, and from these curves, both the nominal inductance LNOM and the winding capacitance CP will be calculated.

Those of you who read through the capacitor section of this chapter will recall that, for the capacitor, the network analyzer is often used to determine the impedance-vs.-frequency characteristics. In the case of the inductor, the network analyzer could also be used to determine the impedance-vs.-frequency characteristics, but the low input impedance of the network analyzer often distorts the true impedance plots. While there are work-arounds for this, replacing the network analyzer with an impedance meter is a much simpler alternative.

The DC ohm meter may be any of the meters commonly available today. The DC ohm meter will be used to measure the coil resistance RS of the model, and, for this reason, a meter which can accurately measure resistance values as low as a tenth of an ohm is desirable.

The model extraction procedure should begin by measuring the coil resistance RS with the ohm meter as shown in Fig. 2.26. The coil resistance may be in the range of a few tenths of an ohm to tens of ohms. Measure the coil resistance, record this value, and set RS of the model to the measured resistance.

After the coil resistance RS has been determined, connect the inductor to the impedance meter as shown in Fig. 2.27. Figure 2.28 illustrates the measured impedance-vs.-frequency characteristics of a typical inductor. At low frequencies, the impedance of the coil remains constant at the value of RS until the frequency F_{RS}. At frequencies less than F_{RS}, the coil resistance determines the impedance of the inductor.

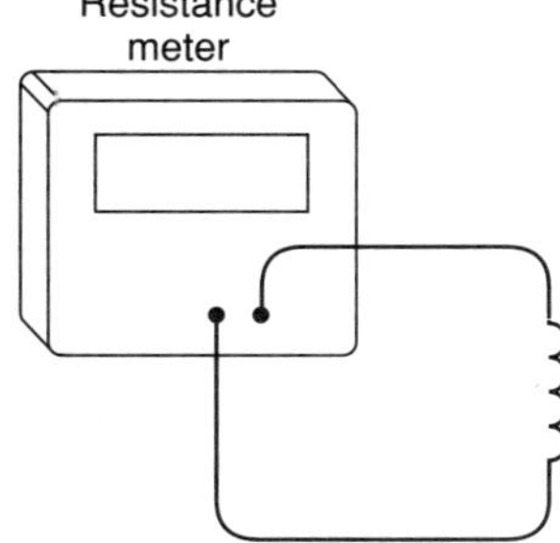

Figure 2.26 Determining the coil resistance of an inductor.

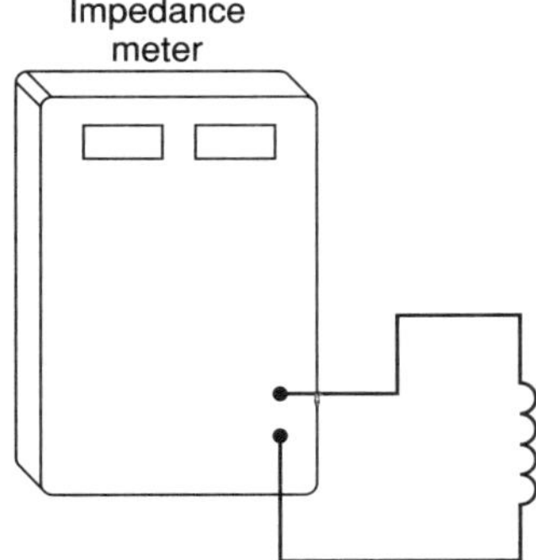

Figure 2.27 Determining the impedance-vs.-frequency characteristics of an inductor.

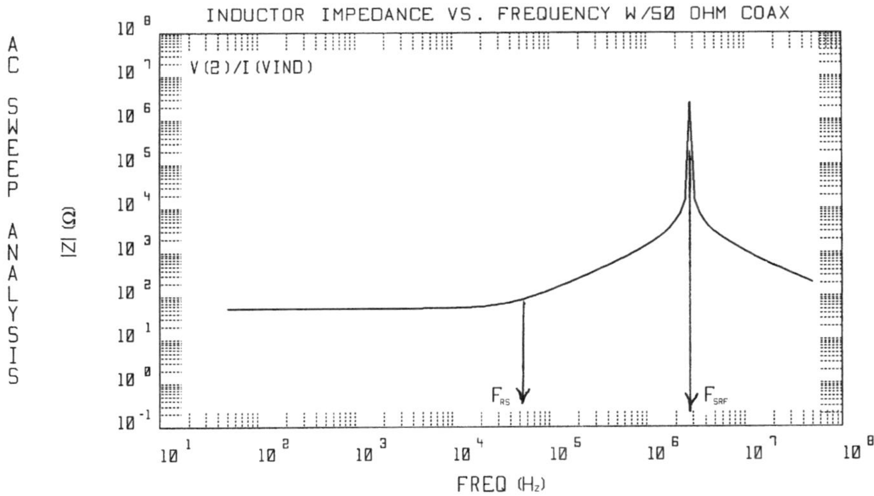

Figure 2.28 Identifying the series resistance F_{RS} and self-resonating F_{SRF} frequencies of an inductor.

Above this frequency, the impedance of the inductor increases with frequency until the coil reaches the self-resonating frequency F_{SRF}. When the measurement frequency reaches the self-resonating frequency, the effect of the coil's winding capacitance becomes apparent and the measured impedance begins to decrease. Determine and record both the F_{RS} and F_{SRF} frequencies.

To determine the nominal inductance LNOM, select a frequency between the F_{RS} and F_{SRF} frequencies and measure the impedance value as shown in Fig. 2.29. Use the measured impedance and Eq. 2.13 to determine the nominal inductance. Record LNOM and set the inductor of the macromodel to this value.

$$\text{LNOM} = \frac{|Z|}{2\pi \text{F}} \tag{2.13}$$

Once the nominal inductance LNOM has been found, use Eq. 2.14 and the F_{SRF} recorded earlier from the impedance-vs.-frequency characteristic to determine the winding capacitance. Set the winding capacitance of the macromodel CP to this value.

$$\text{CP} = \frac{1}{[(2\pi F_{SRF})^2 * \text{LNOM}]} \tag{2.14}$$

With the measurements complete, we need to simulate the macromodel's impedance-vs.-frequency characteristic. In the RCGV34\CH2 subdirectory of the disk accompanying this text is a file named IND.MOD, which contains the inductor macromodel just described.

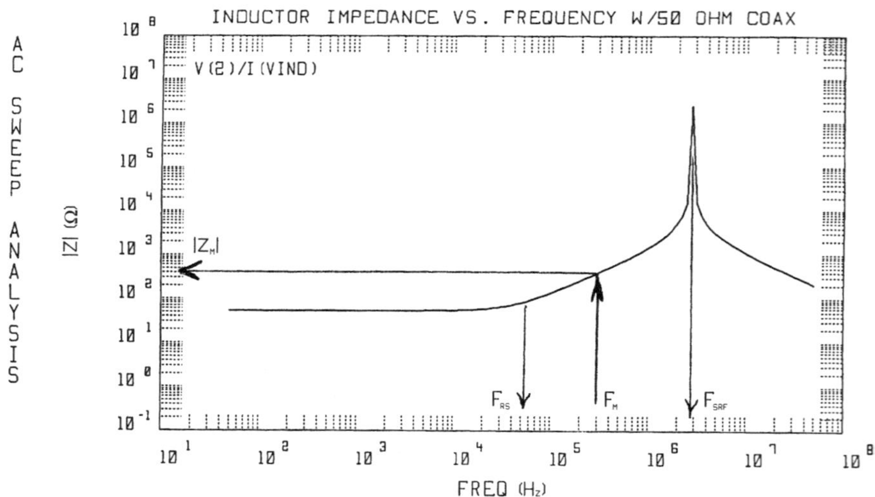

Figure 2.29 Using impedance curves to determine nominal inductance L_{NOM} value.

Also included in the RCGV34\CH2 subdirectory is a SPICE file and a batch file which will simulate the impedance-vs.-frequency characteristics of the inductor. The SPICE file is named IND.CIR and the batch file is named IND.BAT. The batch file simulates the IND.CIR circuit netlist, and the IND.CIR netlist uses the IND.MOD inductor macromodel.

Change into the RCGV34\CH2 subdirectory and edit the LNOM, RS, and CP values of the IND.MOD file so they match the values you just calculated. Once the LNOM, RS, and CP values have been entered, simulate the impedance-vs.-frequency characteristics by typing the command:

```
IND.BAT <enter>
```

Your simulation should be similar to Fig. 2.30.

Using the IND.MOD, IND.CIR, and IND.BAT files, you can quickly produce high-quality, highly accurate inductor macromodels which will simulate the behavior of a real inductor.

Inductor Macromodel Demonstration

To illustrate the value of building inductor macromodels, a 250-μH inductor was selected. Table 2.3 illustrates the measured impedance-vs.-frequency characteristics.

These values were entered into a text file and plotted with the RGRAPH software. Figure 2.31(a) illustrates the resulting plot.

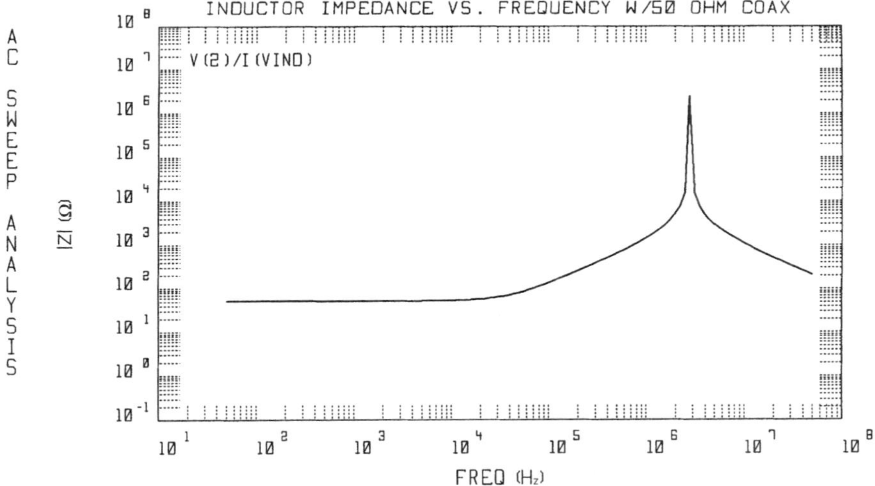

Figure 2.30 Typical inductor impedance characteristics.

TABLE 2.3 Measured Impedance Characteristics of 250-μH Inductor

Frequency	Inductor impedance
10 Hz	.223 Ω
100 Hz	.273 Ω
1000 Hz	1.61 Ω
10 kHz	15.9 Ω
100 kHz	159 Ω
1 MHz	1.67 kΩ
2 MHz	3.87 kΩ
3 MHz	7.91 kΩ
4 MHz	21.0 kΩ
5 MHz	59.0 kΩ
6 MHz	15.3 kΩ
7 MHz	9.27 kΩ
8 MHz	6.80 kΩ
9 MHz	5.43 kΩ
10 MHz	4.56 kΩ

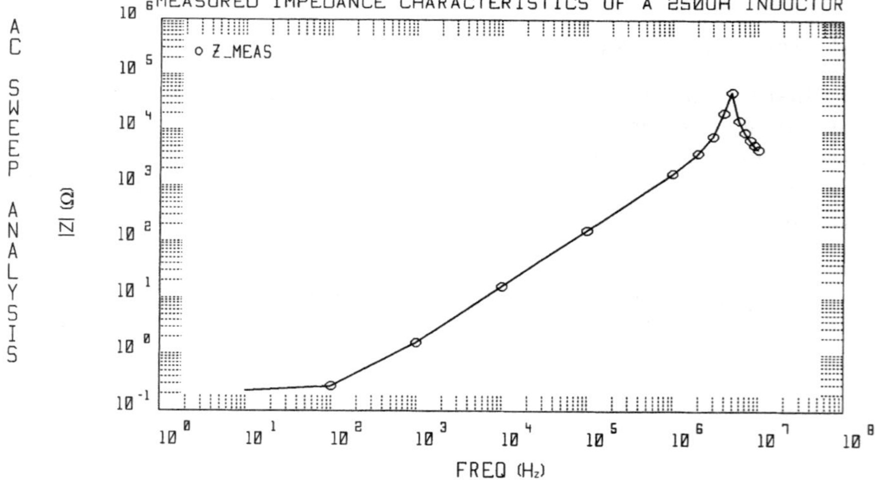

Figure 2.31(*a*) Measured inductor impedance curves.

A DC ohm meter was connected to the inductor and measured .220 Ω of coil resistance, and the RS of the inductor model was set to 0.22 Ω.

At 10 kHz, the measured impedance of the inductor is 15.9 Ω. Using Eq. 2.13 with these values, predict a nominal inductance of 253.1 μH. The LNOM of our macromodel was set to this value.

From inspection of the data, the self-resonating frequency appears to be located somewhere close to 5 MHz. Using 5 MHz and Eq. 2.14 produces a parallel capacitance of 4 pF.

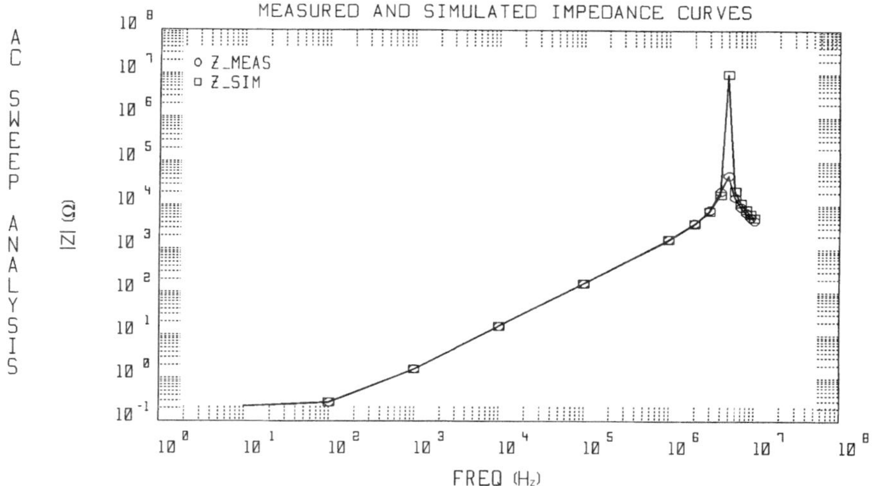

Figure 2.31(*b*) Measured and simulated impedance curves before optimization.

With the model values determined, the IND.MOD file was modified, and simulated with IND.BAT. The result of this simulation, plotted with the original data, is shown in Fig. 2.31(*b*). Using the error function (ERR) described earlier, the simulated model displayed in excess of 63 percent rms error, and most of the error was found at the self-resonating frequency. Looking back at the data, we find that 5 MHz was a poor candidate for the self-resonating frequency. The impedance value at 4 MHz is much higher than at 6 MHz. This indicates that the resonating frequency is actually lower than 5 MHz. Often, as in this case, the data collected may be insufficient to tell the exact self-resonating frequency. However, even without knowing the *exact* resonating frequency, we can still build a good model.

Since the 5-MHz resonant frequency of the model is about that of the actual F_{SRF}, and knowing that LNOM is reasonably accurate, the resonating frequency of the model can be reduced by increasing the parallel capacitance of the model.

Several simulations were performed with different values of CP placed within the inductor macromodel. The values tried were 4.2, 4.4, 4.6, and 4.5 pF. With the capacitance value set to 4.5 pF, the rms error between the two curves fell to 3.88 percent. This simulation is shown in Fig. 2.32(*a*) and with the error between the simulated and measured curves in Fig. 2.32(*b*). The actual error and calculated rms error is shown in Fig. 2.33. Sometimes, to obtain the best fit, we must perform multiple simulations while changing one or more parameters of the model. Often, *manual optimization* will be required to obtain the best-fit model of a device.

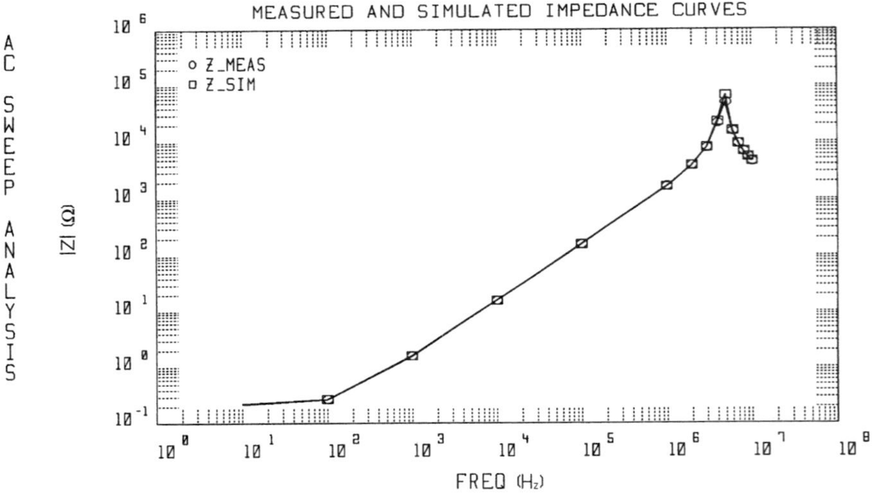

Figure 2.32(*a*) Measured and simulated impedance curves after optimization.

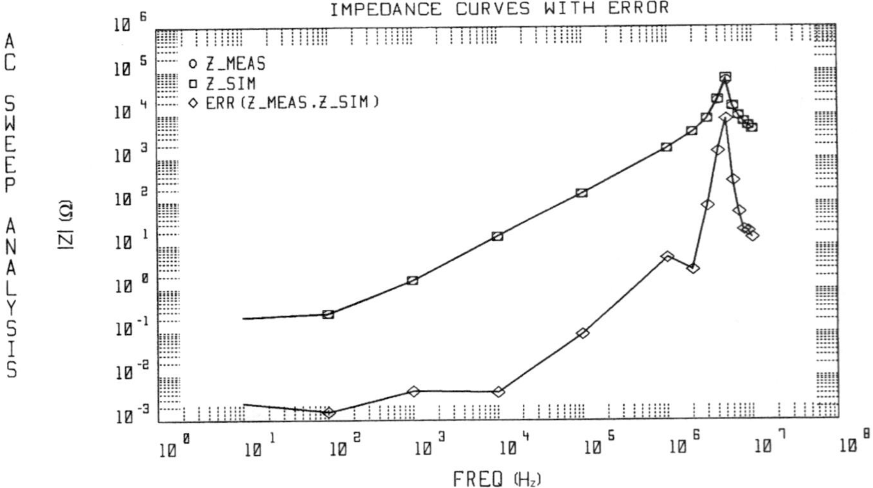

Figure 2.32(*b*) Impedance curves with error.

Once again, with just a few data points, our goal of building a model accurate to within 5 percent rms error took but a matter of minutes to achieve.

Do you need to macromodel all of the inductors in the circuit? Probably not. But inductors used for high-frequency circuits, inductors which are exposed to high-frequency noise, and inductors used in high-power circuits will not be simulated properly without an accurate macromodel.

```
Error report

'Z_MEAS'          'Z_SIM'      Percent Error

22.3000E-02      22.0574E-02        1.1%
27.3000E-02      27.1459E-02        0.6%
16.1000E-01      16.0542E-01        0.3%
15.9000E+00      15.9043E+00        0.0%
15.9000E+01      15.9091E+01        0.1%
16.7000E+02      16.6480E+02        0.3%
38.7700E+02      38.7426E+02        0.1%
79.1000E+02      79.8968E+02        1.0%
21.0000E+03      22.4160E+03        6.7%
59.0000E+03      66.7607E+03       13.2%
15.3000E+03      15.6035E+03        2.0%
92.7000E+02      93.2758E+02        0.6%
68.0000E+02      68.2191E+02        0.3%
54.3000E+02      54.5052E+02        0.4%
45.6000E+02      45.7447E+02        0.3%

RMS Error =     3.88%
```

Figure 2.33 Error report for inductor macromodel.

Inductive Temperature Effects

As with the capacitor, the desire to model temperature-related effects of the inductor macromodel are addressed through the use of a temperature-dependent inductor macromodel. The temperature-dependent inductor macromodel is similar in structure to the original macromodel introduced in Fig. 2.25, but with a temperature-dependent inductor macromodel replacing the LNOM SPICE primitive.

As the temperature of the inductor changes, so does the resistance of the winding. To model this change, the built-in temperature coefficients of the resistor RS can be used. The block diagram for the temperature-dependent inductor macromodel is shown in Fig. 2.34.

At the heart of the temperature-dependent inductor is a voltage-controlled inductor macromodel. (Again, the development of this model is beyond the scope of this chapter but is shown in detail in App. D.) This voltage-controlled inductor has the inductance-voltage relationship shown in Eq. 2.15.

$$L = \text{LNOM} * V_{\text{CONTROL}} \tag{2.15}$$

Equation 2.16 describes the temperature-dependent effects of the inductor.

$$L = \text{LNOM}\,[1 + \text{TC}_1\,(T - T_{\text{NOM}}) + \text{TC}_2\,(T - T_{\text{NOM}})^2] \tag{2.16}$$

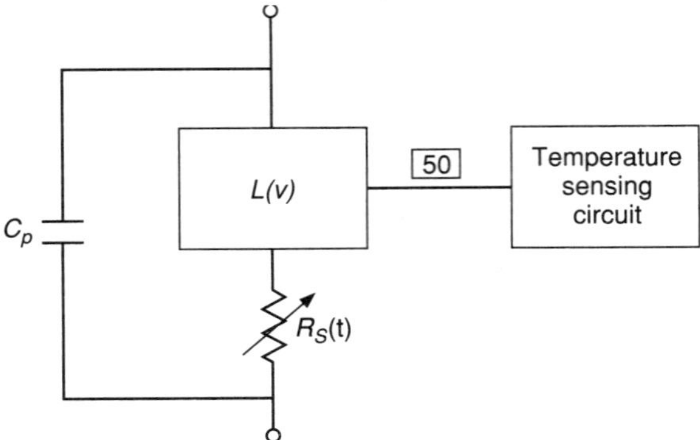

Figure 2.34 Temperature-dependent inductor macromodel.

Borrowing the same temperature-sensing control circuitry used in the temperature-dependent capacitor macromodel, we can easily implement Eq. 2.16. (Those not familiar with the temperature-dependent capacitor macromodel should review this section.) The schematic of the temperature-sensing control circuit is shown in Fig. 2.22. Those of you not familiar with this circuit should take time to read the capacitor section.

The temperature dependence of many inductors can be approximated with a linear and second-order temperature coefficient. For our model, the linear temperature coefficient is assigned through the value of the DC voltage source name VTC1; the second-order coefficient is assigned through the DC voltage source named VTC2. Refer to Table 2.1 for a list of temperature coefficients and the corresponding VTC voltage values. (A program has been written for you to help you determine the VTC1 and VTC2 temperature coefficients of your model. The program is called TC_CALC and is stored in the UTIL subdirectory. Refer to the section of this chapter labeled "Calculating Temperature Coefficients for Passive Components" for instructions on how to use TC_CALC.)

A copy of the temperature-dependent inductor macromodel has been placed in the `RCGV34\CH2` subdirectory. The file is named TIND.MOD. A circuit to simulate the inductance-vs.-frequency characteristics at several different temperatures is also included in this subdirectory. This file is named TIND.CIR. A batch file has been created for you to illustrate the temperature-dependent inductor model. The file is named TIND.BAT. Edit the TIND.MOD model file. Set the temperature coefficient of LNOM with the VTC1 and VTC2 voltage sources and the temperature coefficient of RS with the TC = X statement on the resistor line, then simulate the circuit by typing the command:

TIND.BAT <enter>

Your results should appear similar to those shown in Fig. 2.35.

Careful development of inductor macromodels leads to highly accurate, high-quality simulations. To develop an inductor macromodel, once the measurement equipment has been assembled, takes a matter of minutes. Yet for the small investment in time, the simulated results are immeasurably improved. That's what we call a good return on investment.

Macromodeling Resistors

By now, it should come as no surprise that accurate simulation for many circuits, including high-power and high-frequency circuits, will require more than the simple SPICE resistor primitive. Resistors come in a variety of shapes, sizes, and compositions. Carbon composite, wire-wound, ceramic, and thin- and thick-film all describe different resistor types. With the exception of the wire-wound resistors (wire-wound resistors are best modeled with the inductor macromodel), all of these types can be accurately simulated with the aid of a simple macromodel.

The ideal resistor has a constant resistance or impedance regardless of the frequency of the signal being applied and regardless of the temperature of the resistor. But real resistors change impedance with both frequency and temperature. The task of modeling these behaviors has been simplified for us because the SPICE resistor primitives

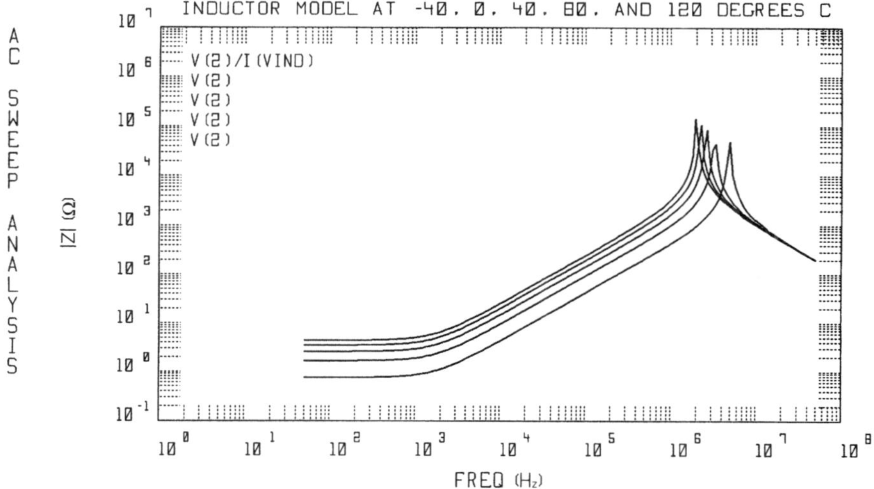

Figure 2.35 Inductor characteristics at five different temperatures.

have user-selectable temperature coefficients built into the device equations. The syntax for the resistor's temperature coefficient is described as follows:

```
RXXXXXX N1 N2 VALUE <TC = TC1, <TC2>>
```

The less-than (<) and greater-than (>) symbols denote an optional portion of the statement. By default, all resistors in SPICE have a temperature coefficient of zero (0 ppm). If the temperature coefficients are set to a nonzero value, the resistance will be adjusted according to the temperature of the circuit and Eq. 2.17.

$$R = \mathrm{RNOM}\,[1 + \mathrm{TC1}\,(T - \mathrm{TNOM}) + \mathrm{TC2}\,(T - \mathrm{TNOM})^2] \quad (2.17)$$

In SPICE, resistors may be described with a linear temperature coefficient (TC1) and with a square-law temperature coefficient (TC2). Either or both of these temperature coefficients may be described on the resistor line. The resistor temperature coefficients may be positive, negative, or zero.

With the question of changing resistance with temperature out of the way, we can focus our attention on the question of changing impedance with frequency. If you observe the impedance-vs.-frequency characteristics of a resistor, you will notice that the impedance begins a steady decrease after a certain frequency. If you plot the impedance vs. the frequency on a log-log axis, you might notice that the curve has the same shape as that of a low-pass filter. Figure 2.36(*a*) illustrates the typical impedance-vs.-frequency characteristic of a 1-kΩ surface-mount resistor.

The impedance of the resistor begins to decrease at high frequencies because of a parasitic parallel capacitance associated with the resistor. This also explains the low-pass filter style characteristics of the impedance curve.

The macromodel we will use for the resistor is shown in Fig. 2.36(*b*). The macromodel consists of the nominal resistance RNOM and a parallel capacitance CP.

Inspecting the several resistor data sheets yields the nominal resistance but reveals very little about the parallel capacitance. Again, the values of RNOM and CP are best extracted from real devices. Extracting RNOM and CP is relatively easy and requires a DC ohm meter or resistive bridge and a capacitance meter.

The ohm meter can be any of those readily available but should have a high-input impedance. Place the resistor across the measurement leads of the ohm meter as shown in Fig. 2.37. Measure and record the nominal resistance value.

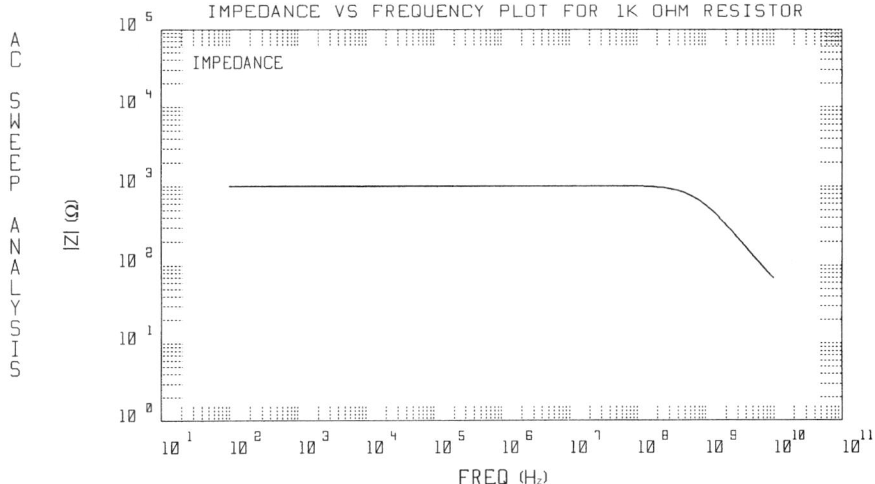

Figure 2.36(*a*) Impedance vs. frequency plot for 1 kΩ resistor.

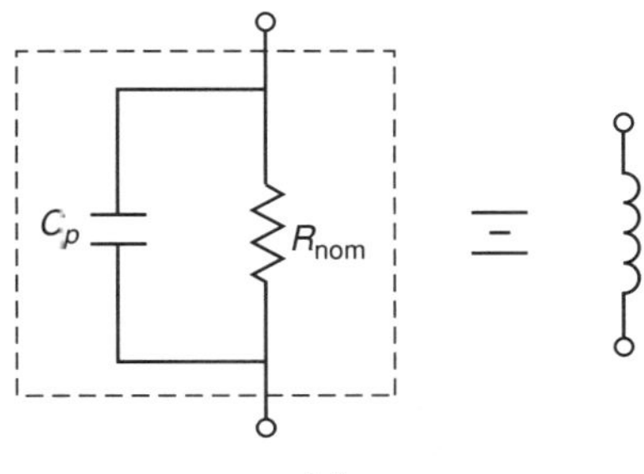

Figure 2.36(*b*) Resistor macromodel.

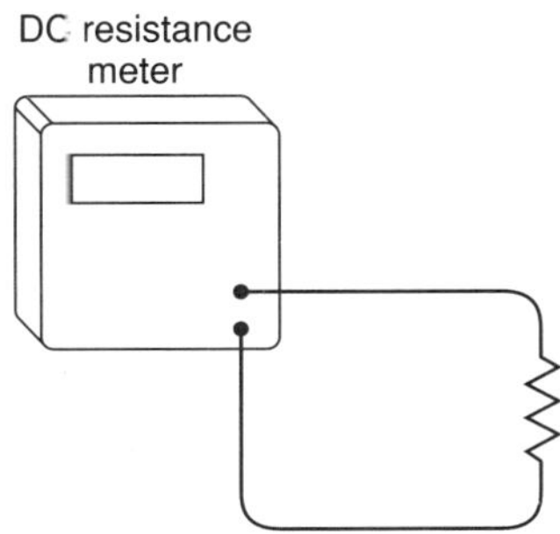

Figure 2.37 Measuring the nominal resistance R_{NOM} for the resistor macromodel.

The capacitance meter may be any of the meters commonly available but should have a means of zeroing or nulling the capacitance of the measurement cabling. Place the resistor across the measurement leads of the capacitance meter as shown in Fig. 2.38. Measure and record the parallel capacitance value.

With the measurements complete, we need to simulate the resulting model. In the `RCGV34\CH2` subdirectory of the disk accompanying this text is a file named RES.MOD, which is a macromodel of a resistor. Also included in this subdirectory is a SPICE circuit file which simulates the impedance-vs.-frequency characteristics of the RES.MOD model. The batch file RES.BAT orchestrates the simulation and plotting for you.

Edit the RES.MOD file and set RNOM and CP to the measured values. Save the file and exit from the editor. Simulate the impedance-vs.-frequency characteristics with the command:

```
RES.BAT
```

The resulting plot should appear similar to that shown in Fig. 2.39.

To verify the quality of the model, you can obtain the true impedance-vs.-frequency characteristics of the device from either a network analyzer (as we did with the capacitor) or from a variable-frequency impedance meter (as we did with the inductor). The impedance-vs.-frequency characteristics of the device should appear very similar to those shown in Fig. 2.39.

Resistor Macromodel Demonstration

To demonstrate the resistor macromodeling technique, a 10-KΩ surface-mount resistor was selected. The nominal resistance and parallel capacitance were determined as previously described. In addition, the resistor was measured with an impedance meter at several different frequencies. This was done so that we could verify the accuracy

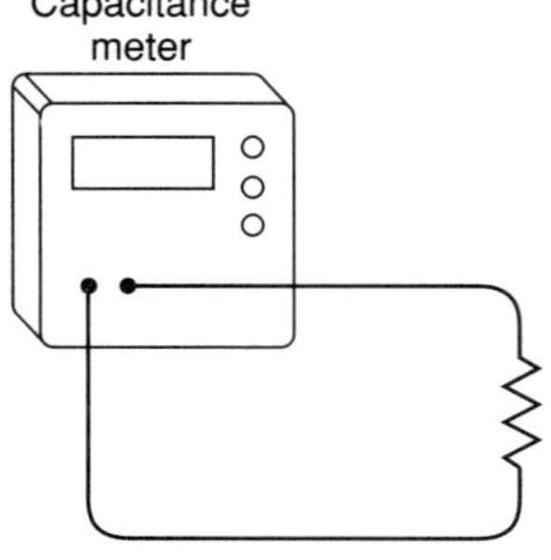

Figure 2.38 Measuring the parallel capacitance C_p of a resistor.

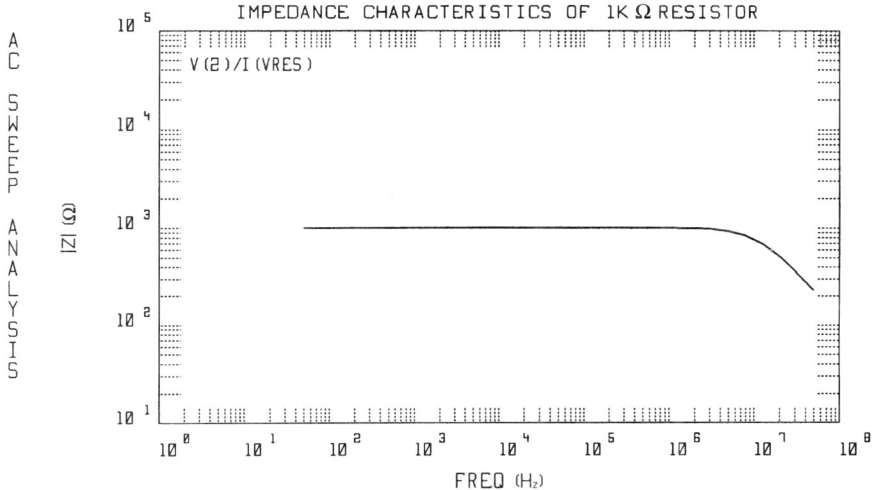

Figure 2.39 Impedance characteristics of surface-mount resistor.

of the model and ensure that we were within our 5 percent rms error tolerance.

Table 2.4 illustrates the measured impedance-vs.-frequency values of the resistor, and Fig. 2.40(*a*) is the RGRAPH plot of this data.

The nominal resistance RNOM was recorded as 10.4 kΩ. The parallel capacitance was measured at 3.4 pF. These values were placed in the RES.MOD macromodel and simulated. Figure 2.40(*b*) shows the measured and simulated impedance curves, and Fig. 2.40(*c*) displays

TABLE 2.4 Measured Impedance-vs.-Frequency Characteristics of 10-K Resistor

Frequency, Hz	Impedance, Ω
10	10400
100	10400
1000	10400
10000	10350
100000	10340
1000000	10120
2000000	9510
3000000	8660
4000000	7780
5000000	6960
6000000	6240
7000000	5630
8000000	5100
9000000	4660
10000000	4270

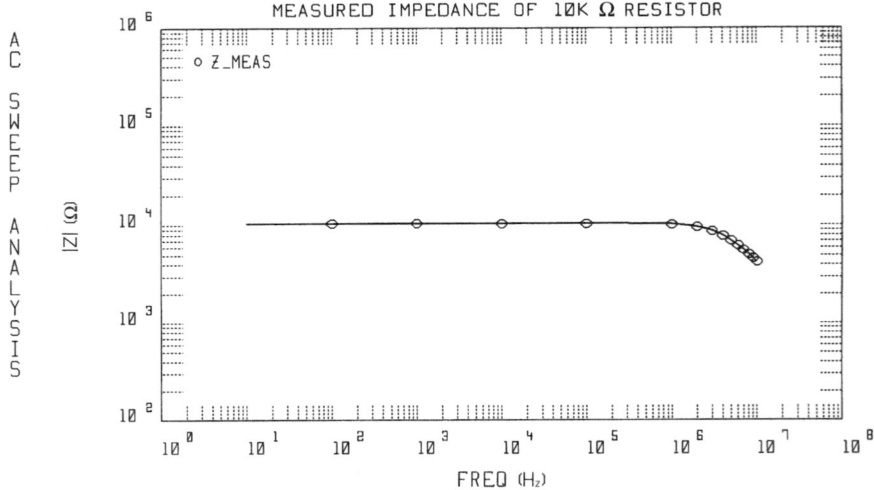

Figure 2.40(*a*) Impedance characteristics of 10-kΩ resistor.

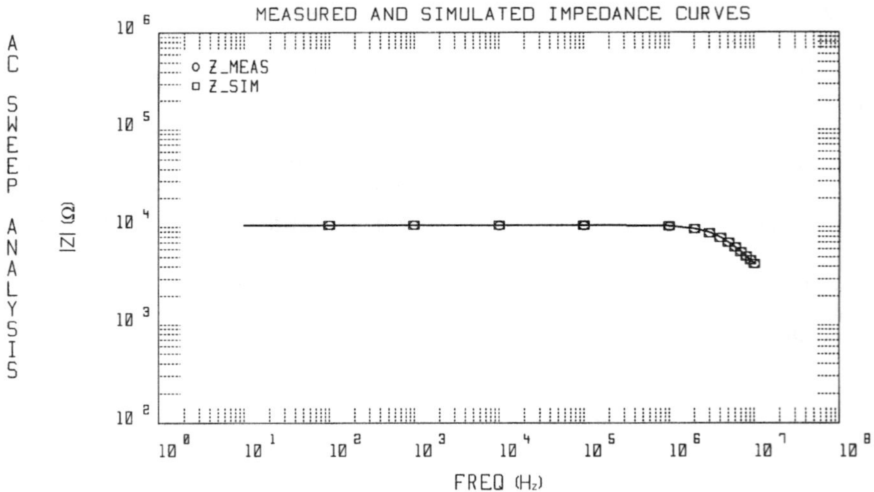

Figure 2.40(*b*) Measured and simulated resistor impedance curves.

the error between the two curves. (You may have to look closely; the error is displayed along the bottom of the x-axis. Impressive, no?) For this device, our model yields a very satisfying .22 percent rms error from 10 Hz to 10 MHz. The error report for this device is shown in Fig. 2.41.

Once again, in a matter of minutes, we have established an accurate model for circuit simulation. While not all simulations require resistor

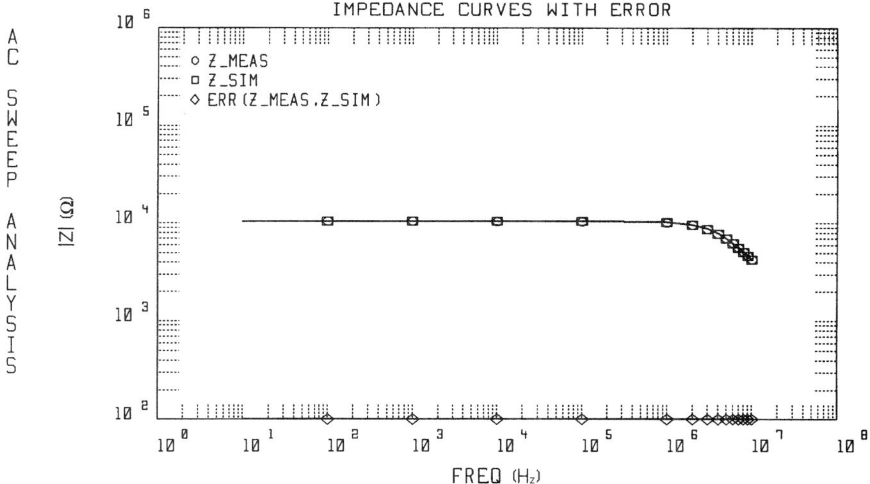

Figure 2.40(*c*) Impedance curves with error.

Error report

'Z_MEAS'	'Z_SIM'	Percent Error
10.4000E+03	10.4000E+03	0.0%
10.4000E+03	10.4000E+03	0.0%
10.4000E+03	10.4000E+03	0.0%
10.3500E+03	10.4000E+03	0.5%
10.3400E+03	10.3974E+03	0.6%
10.1200E+03	10.1525E+03	0.3%
95.1000E+02	95.0398E+02	0.1%
86.6000E+02	86.5391E+02	0.1%
77.8000E+02	77.7381E+02	0.1%
69.6000E+02	69.5808E+02	0.0%
62.4000E+02	62.4088E+02	0.0%
56.3000E+02	56.2475E+02	0.1%
51.0000E+02	50.9957E+02	0.0%
46.6000E+02	46.5184E+02	0.2%
42.7000E+02	42.6857E+02	0.0%

RMS Error = 0.22%

Figure 2.41 Error report for resistor macromodel.

macromodels, circuits which operate at high frequencies will. Of all the passive components, resistors are by far the easiest to model. The time invested in building a resistor macromodel is minimal, yet the results are dramatic. Developing resistor macromodels is an easy means to improving the accuracy and quality of your simulations.

Calculating Temperature Coefficients for Passive Components

The built-in temperature coefficients of the SPICE resistors primitive greatly simplify the task of modeling temperature effects of the resistor. Although the capacitor and inductor primitives are missing these important temperature parameters, our capacitor and inductor macromodels do offer you this option.

To determine the temperature coefficient(s) of the resistor (TC1 and TC2) or to determine the coefficients of the inductor or capacitor (VTC1 and VTC2), we must make several device measurements at different temperatures. This can be done, as shown in Fig. 2.42, by mounting the test device inside a temperature chamber. (Author's note: Experience has shown that home ovens and refrigerators make very poor thermal chambers.) Connect test leads to the test device (resistor, capacitor, or inductor) with a 1-meter or longer cable. Feed the cabling to the outside of the temperature chamber and connect the cables to a meter which will read the nominal value of the device (resistance, capacitance, or inductance).

Once the test device has been mounted in the oven and an appropriate meter has been connected to the device, select several temperatures over the temperature range of interest. Save these temperatures in a list. Before heating or cooling the oven, measure and record the room-temperature nominal device value.

Set the temperature chamber to the first temperature in the list and allow the oven to completely heat-soak the test device. When you are sure the temperature of the device is stable, measure and record the value of the device and the temperature of the oven. Repeat this procedure for each temperature on the list.

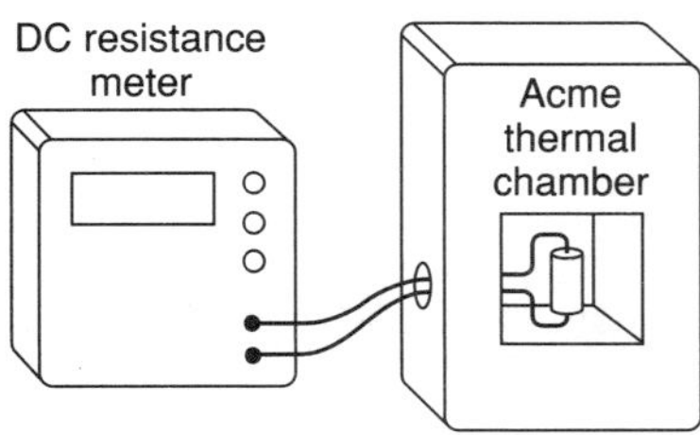

Figure 2.42 Temperature measurements for a resistor.

At this point, a fast and easy method of calculating the temperature coefficients would be handy. With the measured data in hand, change into the `\RCGV34\UTIL` subdirectory. In this subdirectory, you will find two very useful programs. The first is named TC_CALC.EXE, which calculates the first- and second-order temperature coefficients of a resistor, capacitor, or inductor. The program prompts the user for temperature and component values and calculates the temperature coefficients. The second program is called TCB.BAT and it performs the same operations as TC_CALC.EXE but reads the input data from a text file. This is useful for a large number of temperature/value data points. Using these programs is beyond the scope of this chapter; however, the programs are explained in detail in App. A. Use the TC_CALC program to calculate the temperature coefficients of all your passive components and assign these values to the macromodel.

Summary

Often during simulation, little thought is given to the passive components of a circuit. Many circuits, especially high-power, high-current, high-noise type circuits exhibit nonideal behaviors which will not be predicted during simulation by the standard SPICE primitives. To address this problem, simple SPICE macromodels can be developed which will simulate most, if not all, of these nonideal behaviors. All of the macromodels are simple in structure and have element values which are easy to obtain. The time and effort spent in developing these models is rewarded by a highly accurate simulation output.

Chapter

3

Modeling Diodes and Zener Diodes

Introduction

The great football coach Vince Lombardi once said, "The key to success is in learning the fundamentals." In SPICE, the diode model is fundamental to all other semiconductor device models. The two-terminal PN junction diode is the fundamental building block of semiconductor device physics. Many college-level courses on semiconductor devices may spend nearly half of the term on the physics of the diode and the other half dedicated to MOS, bipolar, and JFET devices. The reason for this is simple. The physics of the diode's PN junction is the same as the physics describing the base-emitter, base-collector junctions in a bipolar transistor. These same laws of physics can be applied directly to the drain-substrate, source-substrate junctions of an MOS device. Semiconductor courses spend a large percentage of their class time on the diode because the lessons learned there can be applied to all other semiconductor devices.

In learning to model semiconductor devices in SPICE, the diode model must be described first. The modeling procedures and processes used on the diode, the equations, the parameters, and the measurement and extraction techniques will be applied over and over again as we discuss the other semiconductor devices. Learning to model the diode will provide a good foundation for modeling all the other semiconductor devices in SPICE. You could say that *diodes are fundamental.*

Throughout this chapter, various circuits and models will be described. All of the circuits and models are found in the CH3 subdirectory of the disk that accompanies this text. The utility programs

CJ_CALC, CJB, LINREG, and LRB are found in the UTIL subdirectory of the disk. The disk must be installed on your hard disk and uncompressed before you can access these files.

We will begin our *practical* approach to diode modeling by looking at the standard SPICE diode equations and model parameters. Refer to Table 3.1. The standard model equations are well suited to many types of rectifier diodes and some zener diodes.

However, the equations written into SPICE were a compromise between speed (overall simulation run time) and accuracy. For many applications, the standard diode model equations will not yield acceptable accuracy. Where the model's ability to accurately describe an important characteristic has been compromised, an alternative macromodel will be introduced. The macromodels will help us overcome some of the shortcomings of the standard diode equations.

For those interested in more accurate zener modeling, a macromodel will be used. The zener macromodel uses a standard SPICE diode as the core of the model and adds additional elements to provide a more accurate reverse-biased I-V characteristic. The model can be further extended to describe a temperature-dependent breakdown voltage. The zener macromodel is easily tailored to any breakdown phenomenon including both zener and avalanche effects.

Modeling Approach

In keeping with our philosophy of practical device modeling, typical diode data sheets will be inspected for information which might be used in developing device models. What cannot be obtained from the data sheet will be taken from measured I-V and C-V characteristics. After the initial set of model parameters has been chosen, simulations will be run on the model to optimize the final parameter values and to estimate the accuracy of our model.

The same modeling approach will be used for our macromodels. First, typical data sheets will be inspected for usable modeling information. Then I-V and C-V measurements will be used to extract the remaining parameters and element values. Finally, simulations will be performed to optimize the final parameter values.

Before we begin, however, an introduction to the model parameters is necessary.

Diode Model Parameters

The SPICE diode model is described with 14 model parameters. Of those 14, 5 describe the DC I-V characteristics, 5 describe the capacitive characteristics, 2 describe the built-in temperature variations of

TABLE 3.1 DC I-V Diode Equations

I-V equation	Region of operation
$I_d = \text{IS}^* \left[\exp\left(\frac{qV_d}{\text{N}kt} \right) - 1 \right]$	Forward operation $V_d > -\frac{5\text{N}kt}{q}$ or $V_d >(-0.125 \text{ V at room temperature})$
$I_d = -\text{IS}$	Reverse operation before breakdown $-\text{N}kt \backslash q$ (or -0.125 at room temperature)$>V_d>$BV
$I_d = -\text{IS}^* \left[\exp\left(\frac{(q\text{BV} + V_d)}{kt} \right) - 1 + \text{IBV} \right]$	Breakdown operation Implies $V_d <= \text{BV}$

TABLE 3.2 Diode Model Parameters

Name	Parameter	Units	Default
IS	Reverse leakage current	V	1.0e-14
RS	Diode series resistance	Ω	0.0
N	Emission coefficient	—	1.0
BV	Diode breakdown voltage	V	Infinity
IBV	Diode breakdown current	A	1 mA
CJO	Zero-bias junction capacitance	F	0.0
VJ	Bulk junction potential	V	1.0
FC	Coefficient for capacitance	—	0.5
M	Grading coefficient	—	0.5
TT	Transit time	S	0.0
EG	Energy-band gap	eV	1.11
XTI	Temperature coefficient	—	3.0
KF	Flicker-noise coefficient	—	0.0
AF	Flicker-noise exponent	—	1.0

the device, and 2 are used to describe the flicker-noise generated in the model. These 14 parameters are shown in Table 3.2.

Also shown in Table 3.2, along with the name, is the default value of each parameter. Every model parameter has a default value assigned inside the program. The parameter values found on the .MODEL statements simply override the default values assigned within the program.

The first five parameters IS, RS, N, BV, and IBV, describe the DC I-V characteristics of the diode. Of these five, the first three, IS, RS, and N, describe the forward-bias diode current. BV and IBV describe the after-breakdown, reverse-bias current flow of the device.

The next set of five parameters, CJO, M, VJ, FC, and TT, describe the capacitive characteristics of the model. Of these, CJO, M, VJ, and FC describe the reverse-bias junction capacitance of the diode. TT is used to model the reverse-recovery time of the device.

The parameters EG and XTI model the temperature-dependent changes in diode current and capacitance, while KF and AF model the flicker-noise of the device (KF and AF are used only during an AC noise analysis).

These 14 model parameters completely characterize the diode behavior in SPICE. A practical diode model can be built for almost any rectifier diode from these 14 parameters. The parameters can be used to model junction diodes, GaAs light-emitting and laser diodes, and even Schottky-barrier diodes.

In fact, many accurate device models can be built from something less than all 14 parameters. For example, the parameters KF and AF are used only during an AC noise analysis. If you have no intentions of performing the noise analysis, there is little purpose in taking the time to measure these parameter values. EG and XTI are used only during SPICE temperature sweeps or analyses at different temperatures. If you are not simulating the device over temperature, why bother extracting them? The TT model parameter is used only while switching the diode from forward to reverse bias. For varactor diodes and other diodes which maintain operation in either forward- or reverse-bias, TT has no effect on the model. Thus, many diode models can be built from only a subset of the original 14.

Now to many of you this may sound cavalier, but remember that the goal of this text is to teach you to develop practical device models, and low model development time was one of the criteria of a practical device model.

By themselves, the model parameters do nothing. The model parameters are *scalar variables* of the diode equations in SPICE. To really understand the model parameters and how to extract them, one must first understand the diode equations.

DC-IV equations

The DC I-V characteristics of the diode model are separated into three distinct regions of operation: forward conduction, reverse conduction (before breakdown), and reverse-bias breakdown (if specified). Notice the model parameter BV in Table 3.1. BV is the reverse-bias breakdown voltage of the diode. By default, this parameter has been set to minus infinity, which means if BV is not specified explicitly, the diode will act like a simple rectifier diode and the reverse-bias breakdown equations will not be used. Under these conditions, the diode has only two regions of operation: forward and reverse conduction.

The DC I-V equations follow the classic Shockley diode description.

$$I_d = \text{IS} * \left[\exp\left(\frac{qV_\text{d}}{\text{N}kt}\right) - 1\right] \tag{3.1}$$

In this equation, q is the charge on an electron (1.6022e-19 C); k is Boltzmann's constant (1.3806e-23 J/K); IS is the reverse saturation current; t is the temperature of the diode in degrees Kelvin; and N is the emission coefficient factor. N has a theoretical value of unity but may range from 0.7 to 3 in real diodes.

In Eq. 3.1, two variables are shown in uppercase: IS and N. *Throughout this text, the values shown in uppercase text in the model equations represent the model parameters you must supply for the model.*

The basic exponential relationship is shown in Fig. 3.1. Under forward-bias, the exponentially rising current is shaped by the IS and N parameters. Figure 3.2(a) and (b) illustrate how different values of IS and N affect the diode characteristics.

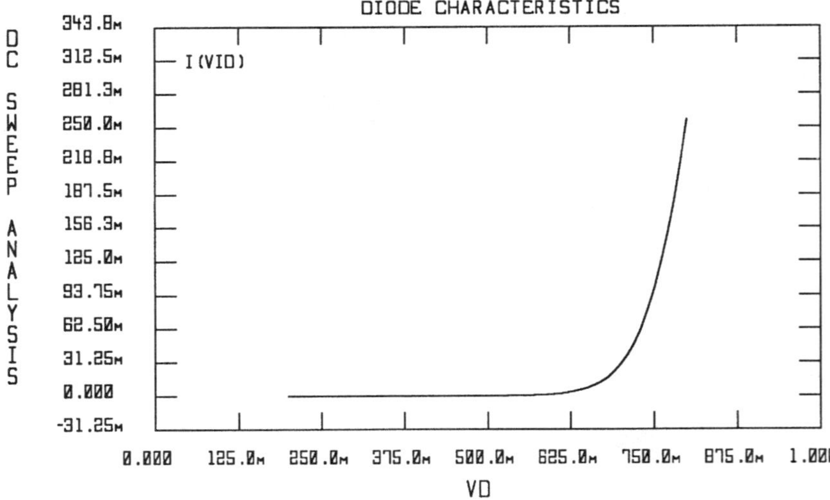

Figure 3.1 Exponential relationship of diode current.

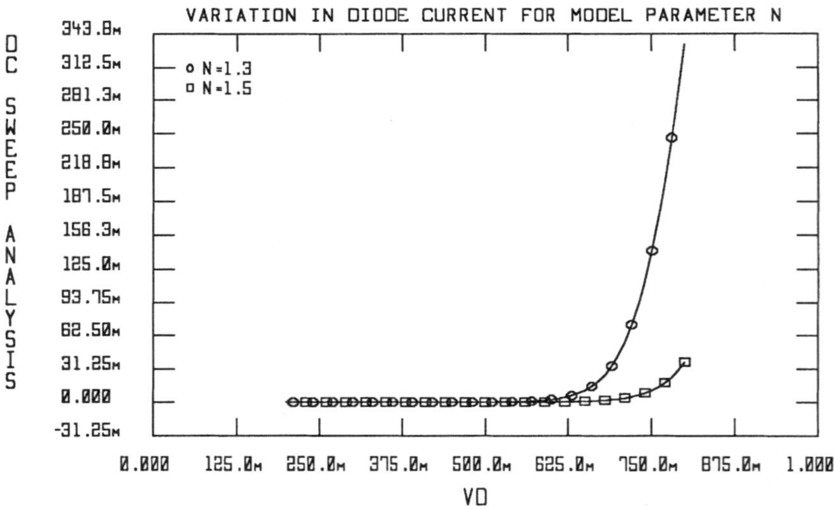

Figure 3.2(*a*) Diode current variations for model parameter N.

While these exponential I-V curves are very common in diode data sheets, it is time to introduce a less common I-V curve which is much more useful in diode modeling. Figure 3.3(*a*) and (*b*) illustrate the forward diode voltage vs. the natural log of the diode current. While Fig. 3.3(*a*) may not look significantly superior to Fig. 3.1, the model parameter IS and N almost jump off of this figure. By comparing Figs. 3.1 and 3.3(*a*) it becomes obvious that, on the linear plot (Fig. 3.1), volt-

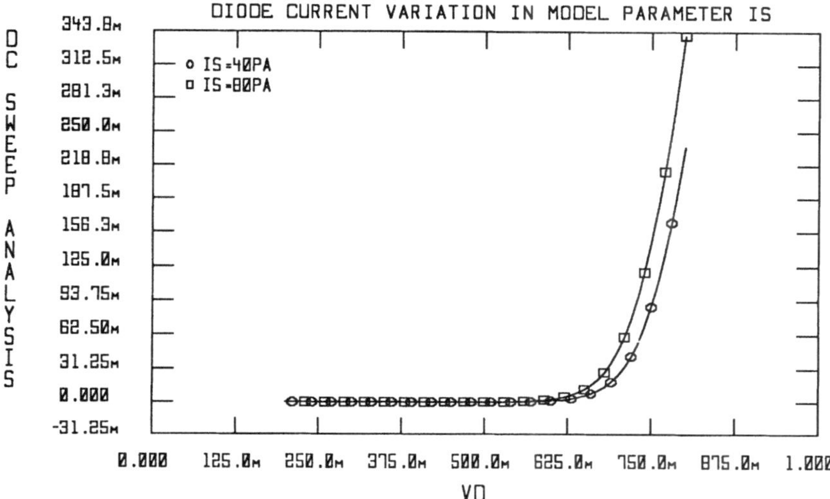

Figure 3.2(*b*) Diode current variations for model parameter IS.

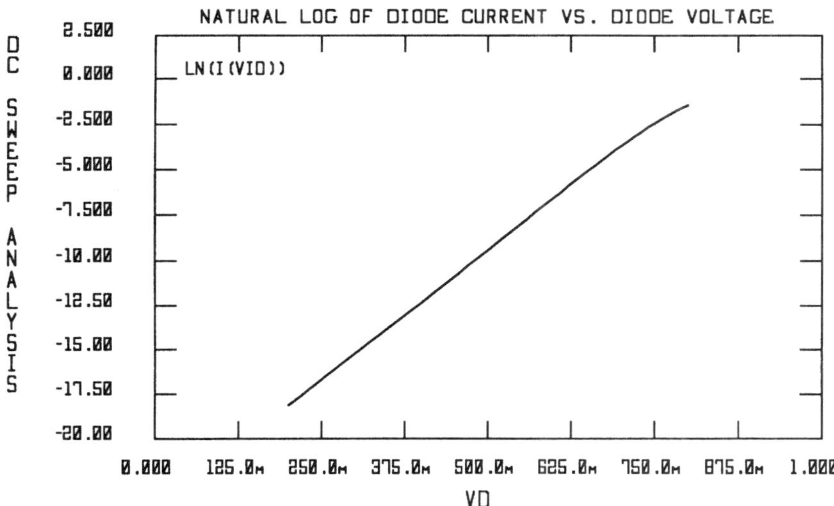

Figure 3.3(*a*) Natural log plot of diode current.

ages less than 0.6 V result in no measurable diode current. This loss of resolution on the linear plot leads designers to believe that diodes don't turn on before 0.6 V. But in reality, diodes follow the same exponential relationship at voltages above 0.6 V as they do at 0.1 or 0.2 V. Unfortunately, these low-voltage current values are often lost due to the poor resolution of the linear plot. Because of the loss of informa-

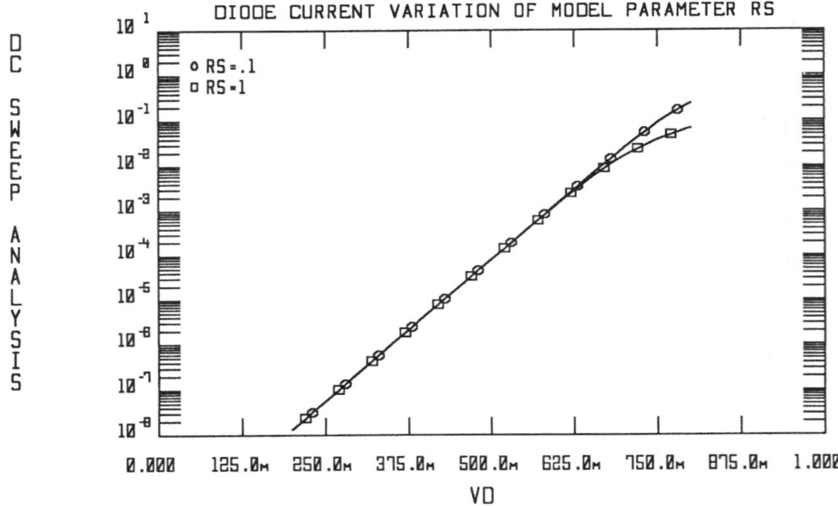

Figure 3.3(*b*) Variation in model parameter RS seen on a log plot.

tion below 0.6 V, for model parameter extraction, we will focus our attention on the natural log diode plot shown in Fig. 3.3(*a*).

Looking back at Eq. 3.1, you might be surprised to find there is no RS parameter in the equation. This is due to the series resistance being modeled as a lumped resistor outside the actual diode model. Figure 3.4 illustrates the equivalent circuit of the diode used in SPICE. Equation 3.1 describes the voltage-current relationship of the dependent current generator. RS represents a fixed-value resistor outside the generator and creates a small voltage drop equal to I_d * RS. To SPICE, the diode voltage is calculated from the points labeled V_d'. This is the voltage you would measure during the simulation if you used a .PRINT or .PLOT statement to observe the diode voltage. Internally, SPICE subtracts the RS voltage drop before applying the

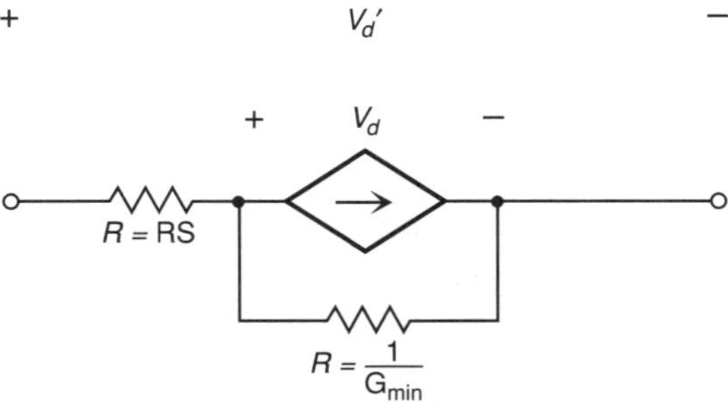

Figure 3.4 SPICE diode equivalent circuit.

diode voltage to the exponential generator. Figure 3.3(*b*) illustrates the effect of the series resistance on the I_d vs. V_d plot.

Figure 3.4 also illustrates the convergence-assist resistor *GMIN*. In my first book, *Inside SPICE: Overcoming the Obstacles of Circuit Simulation*,[1] the function of the *GMIN* resistors is discussed in detail. For those of you who have not read this text, the function of the *GMIN*resistor is to aid convergence. The value of the *GMIN* resistor should be set for the convergence characteristics of your circuit, not to increase the accuracy of a particular model. Remember, the value of *GMIN* is a global parameter. Changing the value of *GMIN* will reset all the *GMIN* resistors on all semiconductor devices. Since the default value of *GMIN* is 1e-12 mhos, an equivalent resistance of 1000 GΩ, this element will have very little influence on the I-V relationship of the model.

Once the diode becomes reverse-biased, the exponential expression of Eq. 3.1 begins to diminish quickly. In fact, after a diode voltage of about −125 mV, the exponential expression is very small when compared with the value of unity being subtracted from it. For all practical purposes, below about −125 mV, Eq. 3.1 can be rewritten as shown in Eq. 3.2. The authors of SPICE recognized this too. In the diode equations, once the diode voltage drops below $-5 * V_t$ (where V_t is the classic thermal voltage given by Eq. 3.3), the exponential expression is completely dropped from the equations. Below this voltage, the diode current remains fixed at −IS.

$$I_d = -\mathrm{IS} \tag{3.2}$$

$$V_t = \frac{kt}{q} = 25.5 \text{ mV at room temperature} \tag{3.3}$$

If the diode model parameter BV has been specified and the diode voltage continues dropping below the breakdown voltage, the diode current will again begin following an exponential characteristic, but this time, the current increases in the negative direction. When the diode voltage rests at exactly the breakdown voltage, the diode current will be described by the value of IBV. If the diode continues into reverse bias, the reverse-bias breakdown current will increase exponentially as expressed in Eq. 3.4.

$$I_d = \mathrm{IS}\,[\exp(q*(-\mathrm{BV}' + V_d/kt - 1] - \mathrm{IBV} \tag{3.4}$$

Notice that BV has been replaced with the parameter BV′. SPICE calculates the pseudo-breakdown voltage to help reduce nonconvergence. If users were allowed to specify a constant reverse-biased current (IS) and a different breakdown current value (IBV), a large discontinuity would exist just at the point of diode breakdown. Since discontinuities in the model equations cause nonconvergence problems (see Chap. 3 of *Inside SPICE: Overcoming the Obstacles of Circuit Simulation*[1]), the authors of SPICE added a mechanism to help minimize the size of any discontinuity. When a user specifies a breakdown voltage, SPICE calculates the pseudo-breakdown voltage (BV′) so that (1) the reverse-biased breakdown region current and the before-breakdown reverse-saturation current (IS) exhibit a smooth transition from one region to the other, and (2) at exactly the breakdown voltage specified (BV), the diode is sourcing the proper breakdown current (IBV). The pseudo-breakdown voltage is something you never see. But by implementing the equations as they did, the authors of SPICE improved the convergence characteristics of the program and made extracting the breakdown parameters very easy.

For our diode models, the breakdown voltage BV parameter will be determined from the point of maximum curvature on the reverse-bias I-V plot, and the reverse-bias breakdown current IBV will be the corresponding current at the breakdown voltage BV.

Equations 3.1, 3.2, and 3.4 describe the three regions of diode operation. These three regions are summarized in Table 3.3.

In the forward region, Eq. 3.1, the IS, N, and RS parameters accurately model the forward conduction of many types of diodes. But notice that the IS model parameter used to describe the forward characteristics is also used to model the reverse-leakage current. This is one of the common problems with the SPICE diode model. The reverse-bias current is not actually constant as the SPICE equations would predict. The reverse-bias current of real diodes is actually exponential in shape (albeit the exponential is very, very soft). But for years, people have modeled the reverse-bias diode current as a constant value. We will model the reverse-bias current both as a constant and as an exponential. We will begin to model the reverse-leakage current with the simple IS model parameter and later with a macromodel to improve the accuracy of the simulation model. For many circuits, the IS model parameter will be accurate enough to meet our

TABLE 3.3 SPICE Diode Equations for All Regions of Operation

$I_d = \mathrm{IS}\left[\exp\left(\dfrac{q\,V_d}{\mathrm{N}kt}\right) - 1\right]$	for	$V_d \geq \dfrac{5\,\mathrm{N}kt}{q}$
$I_d = -\,\mathrm{IS}$	for	$-\mathrm{BV} < V_d < \dfrac{-5\mathrm{N}kt}{q}$
$I_d = -\,\mathrm{IS}\left[\exp\left(\dfrac{-q\,(\mathrm{BV}' + V_d)}{kt}\right) - 1\right] - \mathrm{IBV}$	for	$V_d < -\,\mathrm{BV}$

goal of 5 percent rms accuracy. But for those willing to take the extra time, we can improve the diode characteristics even further with the macromodel.

Another interesting feature (or shortcoming) of the diode model becomes obvious when you start to model zener and avalanche diodes. Inspecting Eq. 3.4, you may notice that the reverse-bias exponential expression does not have the emission coefficient N parameter of the forward exponential. The absence of this parameter means that the shape of the reverse-bias breakdown characteristic is fixed and cannot be changed. For designers who use zener diodes and other clamping diodes, this may present an unacceptable situation. As before, to remedy this situation, a macromodel of the diode will be introduced which allows you to tailor the reverse-breakdown characteristic with all the flexibility of the forward exponential.

DC capacitance equations

The DC capacitance characteristics refer specifically to the junction capacitance of the diode. When a P-type semiconductor and an N-type semiconductor are brought together, a depletion region forms between the two, as shown in Fig. 3.5. The depletion region is an area void of any charge carriers and, as such, acts like an insulator between the N- and P-type regions of the diode. Because the N-type and P-type regions of the device are loaded with mobile carriers, and because the depletion region separates these carriers, a junction capacitance is formed.

The width of the region is determined by the doping concentration of the materials and the voltage bias applied to the diode. If a capacitance meter is placed across a diode, the junction capacitance can easily be measured. Equation 3.5 shows the junction capacitance. In this equation, C_j is the junction capacitance, E_{si} is the dielectric permittivity of silicon, A is the area of the junction, and X_j is the thickness of the depletion layer.

$$C_j = E_{si} * \frac{A}{X_j} \tag{3.5}$$

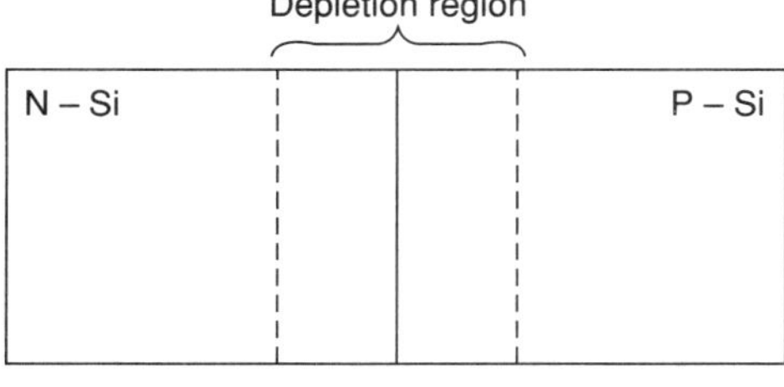

Figure 3.5 Depletion region of a diode.

Under forward bias, the depletion region edges move closer together and the measured junction capacitance increases—at least until the diode begins to conduct a significant amount of current and the capacitance is effectively shorted. Under reverse bias, the depletion region edges move apart and the junction capacitance decreases. This effect is illustrated in Fig. 3.6(*a*) and (*b*).

In SPICE, the junction capacitance is modeled by the model parameters CJO, VJ, M, and, to a lesser extent, FC. The junction capacitance relationship is shown in Eq. 3.6. CJO is the zero-biased junction capacitance and is the measured capacitance with no voltage applied to the diode. (CJO can be measured with a simple capacitance meter.) VJ represents the built-in potential of the diode, and M represents the grading coefficient of the device. Both M and VJ are related to the way the diode is fabricated. VJ represents the theoretical built-in potential of the diode.

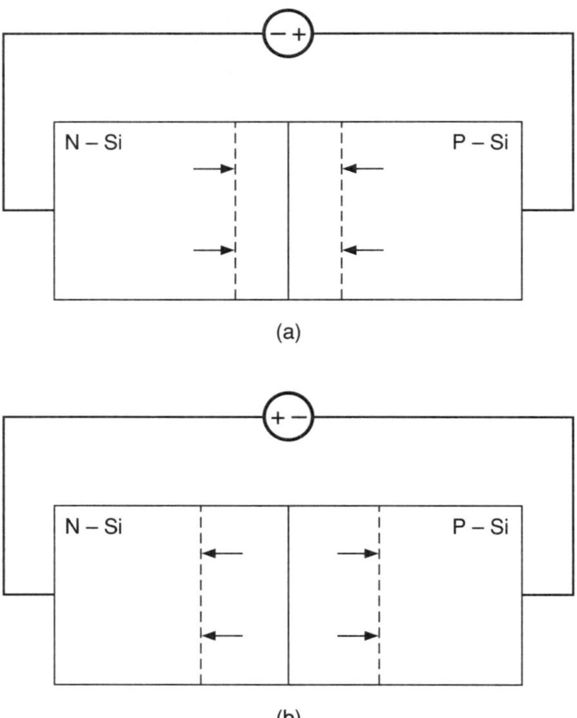

Figure 3.6(*a*) Depletion region change under forward bias; (*b*) depletion region change under reverse bias.

$$C_j = \frac{\text{CJO}}{\left[1 - \left(\frac{V_d}{\text{VJ}}\right)\right]^{\text{M}}} \tag{3.6}$$

When forward biased, the depletion region edges move closer together. If the bias is increased, the edges move even closer together. In theory, this should continue until the point where the depletion region edges eventually touch. If this ever happened, XJ in Eq. 3.5 would be zero, and the junction capacitance at this voltage would be infinite. In real devices, this never happens because forward-bias current flow keeps the depletion regions separated.

In SPICE, VJ represents the forward-biased voltage at which the depletion region width theoretically shrinks to zero. This parameter has a large influence on the capacitance-voltage characteristics under forward- and slightly reverse-biased conditions.

The grading coefficient M is a function of the type of processing done to fabricate the diode and the doping concentrations of the N- and P-type regions. M has a pronounced effect on the entire junction capacitance curve.

Figure 3.7(*a*), (*b*), and (*c*) illustrate the influence of CJO, VJ, and M on the capacitance characteristic.

The parameter FC is a coefficient to help overcome numerical difficulties if and when the diode voltage (V_d) is equal to VJ. Under these conditions, solving Eq. 3.6 would result in a divide-by-zero error and

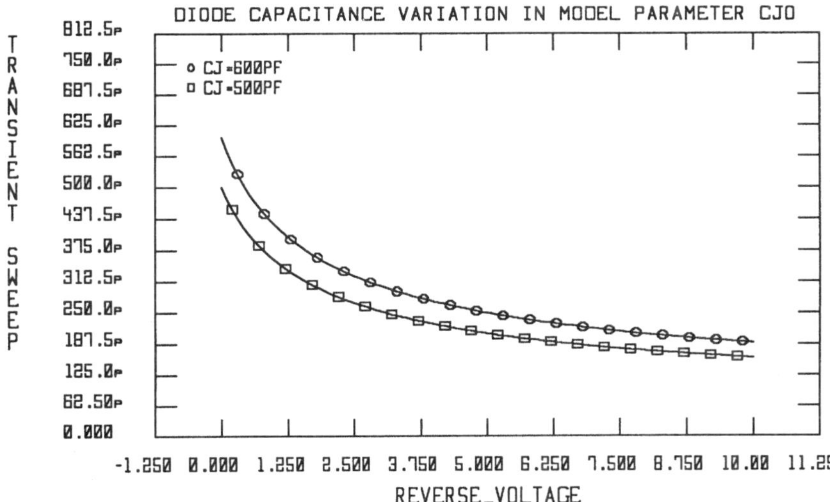

Figure 3.7(*a*) Diode capacitance variation for model parameter CJO.

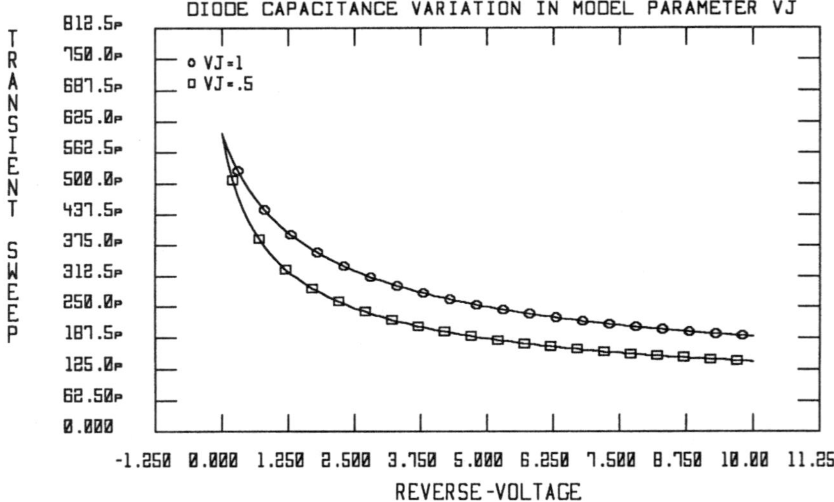

Figure 3.7(*b*) Diode capacitance variation for model parameter VJ.

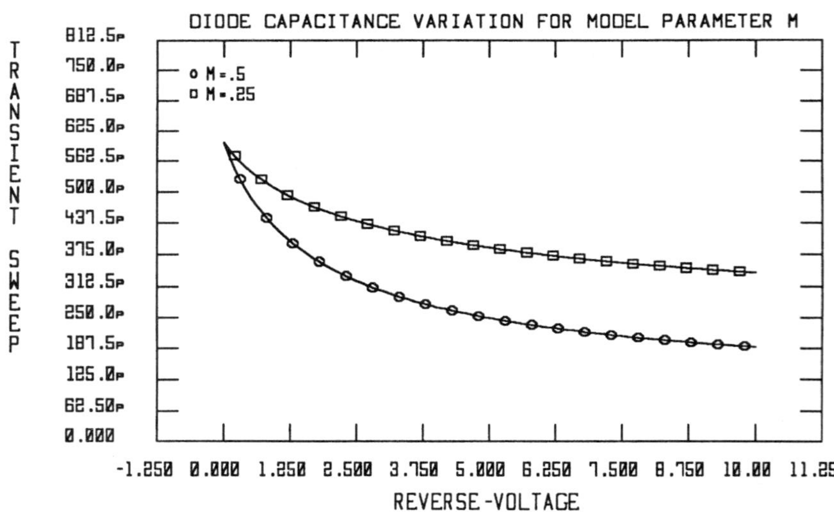

Figure 3.7(*c*) Diode capacitance variation for model parameter M.

crash the program. Even if the diode voltage was only close to VJ, the expression could easily cause an overflow error.

To avoid numerical problems, the authors of SPICE modified the equations of the junction capacitance for forward-biased voltages and introduced the parameter FC. FC has a default value of 0.5. When the diode voltage reaches a value equal to FC * VJ, the capacitance expression shown in Eq. 3.6 changes to the expression shown in Eq.

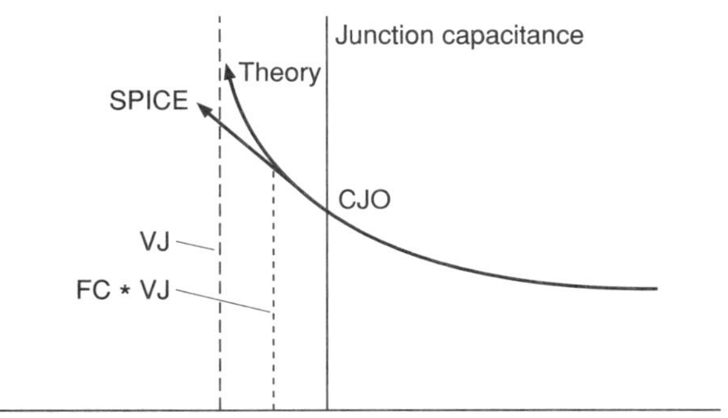

Figure 3.8 The theoretical junction capacitance vs. the capacitance model in SPICE.

3.7 and the diode's capacitance increases linearly with increasing voltage rather than hyperbolically. The difference between Eqs. 3.6 and 3.7 is shown graphically in Fig. 3.8.

$$C_j = \frac{\mathrm{CJO}}{(1 - \mathrm{FC})^{1 + \mathrm{M}}} \left[1 - \mathrm{FC}(1 + \mathrm{M}) + \frac{\mathrm{M}^* V_d}{\mathrm{VJ}} \right] \tag{3.7}$$

While the numerical algorithms in SPICE require the FC coefficient to avoid difficulties, the addition of this parameter only degrades the accuracy of the capacitance model. By adding FC to the equations, the capacitance model for voltages greater than FC*VJ is compromised. For an accurate diode model, FC should be set as high as possible without inducing numerical problems. *In this text, FC will be set to the default value of 0.5 for all our models.*

Transient capacitance equations

While the junction capacitance is a function of the DC bias of the diode, there is a second capacitance associated with any PN junction which is only observable under transient (switching) conditions.

While forward-biased, N-type carriers from the N-type silicon are injected into the depletion region. Because of the high field strength in the region, these carriers are quickly swept over the P-type side of the diode. N-type carriers in a P-type silicon will begin to recombine at defect sites in the crystal. As more and more forward bias is applied to the device, more and more N-type carriers are forced into the P-type side of the diode. Soon a surplus or excess of N-type carriers exists in the P-type side of the device. These excess carriers must "wait around" until a recombination site becomes available in the P-doped crystal. At the same time, the same action is occurring in the

N-type side of the diode. P-type carriers are waiting for recombination sites in the N-type silicon.

Under forward-bias conditions, the excess carriers in the device are not a problem; however, if the diode is quickly switched from forward to reverse bias, the carriers which are trapped on opposite sides of the junction still recombine at the same rate as they did under forward bias. But now the diode is reverse-biased and the current actually flows backward across the depletion region and appears as a large transient reverse-leakage current. This large surge in current occurs for only a short period of time, until the excess carriers in the P- and N-type sides are exhausted. This effect is known as the *stored-charge capacitance* and is modeled in SPICE with the TT model parameter. Equation 3.8 is used in SPICE to model the stored-charge capacitance.

$$C_j = \mathrm{TT} * \frac{dI_d}{dV_d} \tag{3.8}$$

Temperature equations

In the diode model, the IS and CJO values of the model are modified over temperature as shown in Eqs. 3.9 and 3.10. The temperature coefficients XTI and EG shown in these equations affect only the temperature characteristics of the device. When simulations are run at the nominal temperature (27 C), XTI and EG have no impact on the model.

$$\mathrm{IS}(T) = \mathrm{IS}(T_{\mathrm{nom}}) * \left(\frac{T}{T_{\mathrm{nom}}}\right) ** \left(\frac{\mathrm{XTI}}{\mathrm{N}}\right) * \exp\left(q * \mathrm{EG} * \frac{(T - T_{\mathrm{nom}})}{(k * \mathrm{N} * T * T_{\mathrm{nom}})}\right) \tag{3.9}$$

$$\mathrm{CJO}(T) = \mathrm{CJO}(T_{\mathrm{nom}}) *$$

$$(1 + \mathrm{M} * (400\mathrm{E}\text{-}6 * (T - T_{\mathrm{nom}}) - \left(\frac{(\mathrm{PHI}(T) - \mathrm{PHI}(T_{\mathrm{nom}}))}{\mathrm{PHI}(T_{\mathrm{nom}})}\right) \tag{3.10}$$

The EG parameter describes the energy band-gap of the semiconductor. The band-gap will change depending on which type of semiconductor is used to fabricate the device. The band-gap values can be found in many books covering semiconductor physics including Ref. 2.

Noise equations

The noise parameters KF and AF are used only during an AC noise analysis. Since these parameters have no influence on the normal

DC, AC, or transient response of the circuit, we will present no further discussion on them.

Extraction Procedures

I-V parameters

The extraction procedures for the diode begin with what are probably the most important of the diode parameters, IS, RS, and N. To extract these parameters, a graph of the natural log of I_d vs. the forward diode voltage is necessary. This plot might come from a manufacturer's data sheet or from measured devices. A typical diode plot is shown in Fig. 3.9.

If you plan on measuring real diodes, you might investigate a parametric tester. A parametric tester is a fancy curve tracer. It has multiple source-measure units, and these meters can measure current while forcing a voltage or measure a voltage while forcing a current. Most parametric testers allow you to plot data on linear or logarithmic axes. Many testers allow you to make pulse measurements to help eliminate device heating. Some testers even allow you to do curve-fitting on the measured data. For model parameter extractions, a good parametric tester can greatly ease the extraction process.

To generate I-V plots from real diodes, use the test setup shown in Fig. 3.10. Be careful about driving too much DC current through the diode. If the diode starts to get hot, the measurement will be invalid. For high-current diodes, you may need to pulse the measurement voltage to keep from heating the device.

The IS and N model parameters are easily obtained from the plot of natural log of I_d vs. diode voltage shown in Fig. 3.9. To determine these parameters, simply draw a straight line through the lower portion of the curve as shown in Fig. 3.11. The intersection of the straight-line approximation of the y-axis is the natural log of the IS model parameter, and the slope of the line will be equal to $q/\mathrm{N}kt$, where q is the charge on an electron, k is Boltzmans constant, and t is

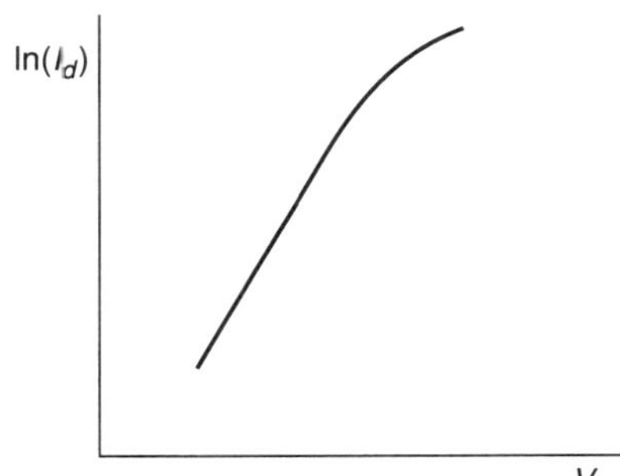

Figure 3.9 Plot of ln (I_d) vs. diode voltage.

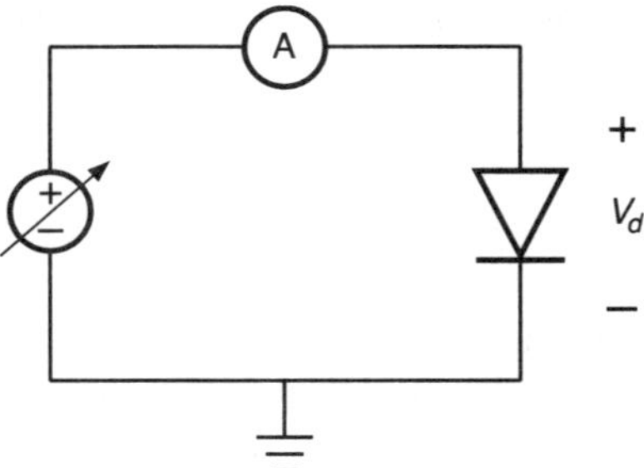

Figure 3.10 Test setup for diode I-V characterization.

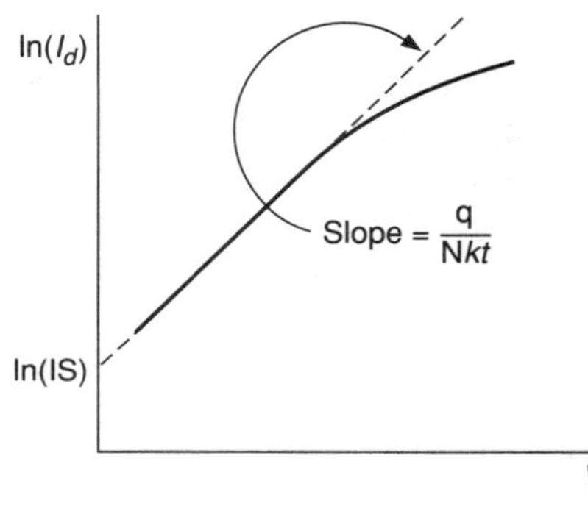

Figure 3.11 Linear approximation to diode characteristics.

the temperature of the diode. If you prefer a more accurate model, use linear regression to determine the slope and intersect of the linear portion of the I-V plot. Equation 3.11 can be used to calculate the value of IS from the *y*-intercept. Equation 3.12 can be used to calculate the N parameter from the slope of the line. If the measurement data is taken at room temperature, Eq. 3.13 can be used to calculate N.

$$\text{IS} = \exp(y\text{-intercept}) \tag{3.11}$$

$$\text{N} = \frac{1}{(\text{slope}*(kt/q)} \tag{3.12}$$

$$\text{N} = \frac{1}{(\text{slope}*0.0255)} \tag{3.13}$$

To calculate RS, draw a horizontal line from the highest measurable diode current to the *y*-axis as shown in Fig. 3.12. The intersection of this line and the *y*-axis can be used to determine the maximum current through the diode, I_{max}. The line you draw represents the I_{max} level of current for both the measured diode curve and the ideal, straight-line curve. Next, measure the voltage drop from the ideal, straight-line approximation to the measured curve as shown in

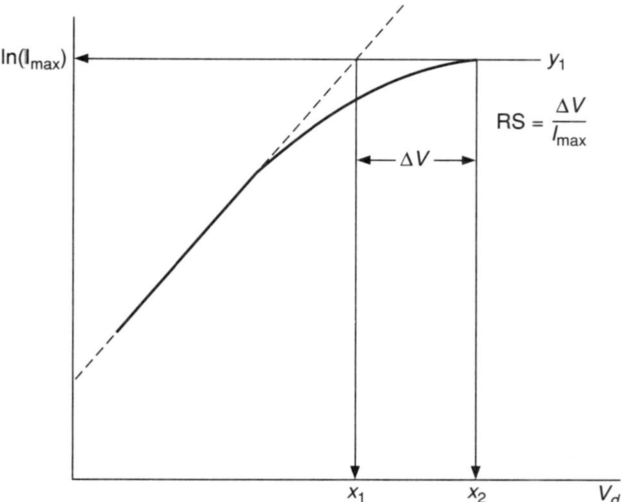

Figure 3.12 Calculating the RS model parameter.

Fig. 3.12. The series resistance model parameter RS can be determined graphically from Fig. 3.12 or from Eqs. 3.14a–d and the slope and intercept of the ideal (straight-line) diode current curve.

$$\mathrm{RS} = \frac{(x_2 - x_1)}{I_{\mathrm{max}}} \tag{3.14a}$$

$$x_1 = \left(y_1 - \frac{(y\text{-intercept})}{\text{slope}}\right) \tag{3.14b}$$

$$x_2 = V_d \mathbin{@} I_{\mathrm{max}} \tag{3.14c}$$

$$y_1 = \ln\,(I_{\mathrm{max}}) \tag{3.14d}$$

To extract the BV and IBV parameters, the reverse I-V characteristics are required, as shown in Fig. 3.13(*a*). Again, this plot can come from a manufacturer's data sheet or from measurements taken from your devices. To determine BV and IBV, simply determine the reverse-bias voltage where the diode begins to break down. This voltage is equal to −BV. Once the BV voltage has been determined, measure the diode current at precisely the breakdown voltage. This current is equal to −IBV. These measurements are shown in Fig. 3.13(*b*). Notice that both of these measured values represent the negative value of the parameter, because *both BV and IBV are entered as positive model parameters.*

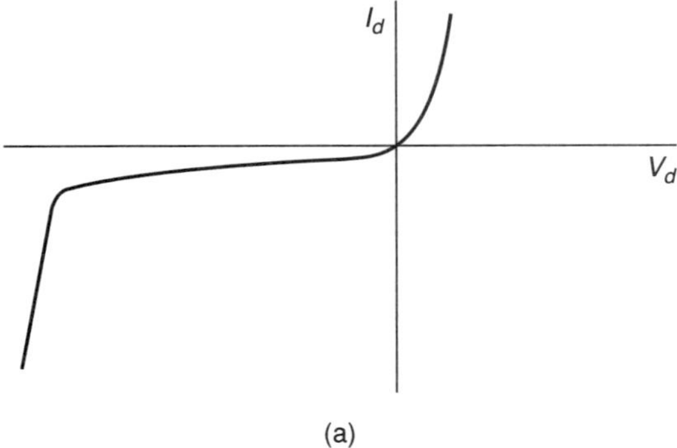

Figure 3.13(*a*) Reverse I-V diode characteristics.

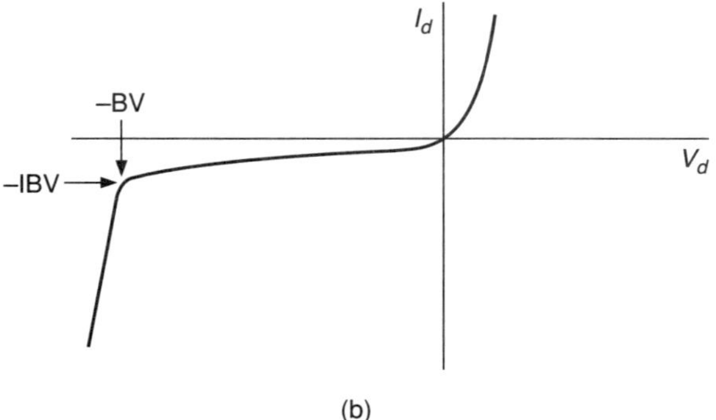

Figure 3.13(*b*) Determining the BV and IBV model parameters.

Capacitance parameters

To determine the capacitance parameters CJO, M, and VJ, the reverse-biased capacitance-vs.-voltage characteristics are required. A typical diode C-V plot is shown in Fig. 3.14(*a*). To measure the capacitance-voltage characteristics requires a capacitance meter which allows you to measure the device capacitance while changing the voltage bias being applied to the diode. Both Hewlett-Packard and Keithly Instruments make this type of meter.

CJO is easily obtained from the intersection of the curve onto the *y*-axis, as shown in Fig. 3.14(*b*). The values of M and VJ are slightly more difficult to obtain than CJO. To obtain M and VJ, the C-V data must be plotted on a graph different from the ones shown in Figs. 3.14(*a*) and (*b*). The graph we need to use is shown in Fig. 3.15(*a*). Notice the *x*-axis variable in Fig. 3.15(*a*) is the $\log(1 - V_d/\mathrm{VJ})$ and the *y*-axis variable is $\log(C_j/\mathrm{CJO})$, where C_j is the measured capacitance

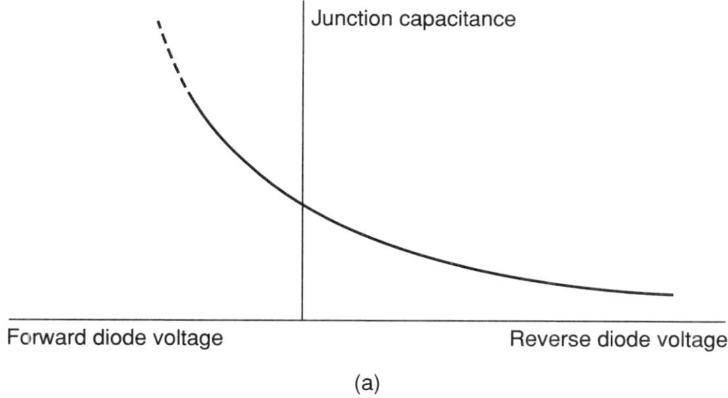

Figure 3.14(*a*) Typical junction capacitance characteristics.

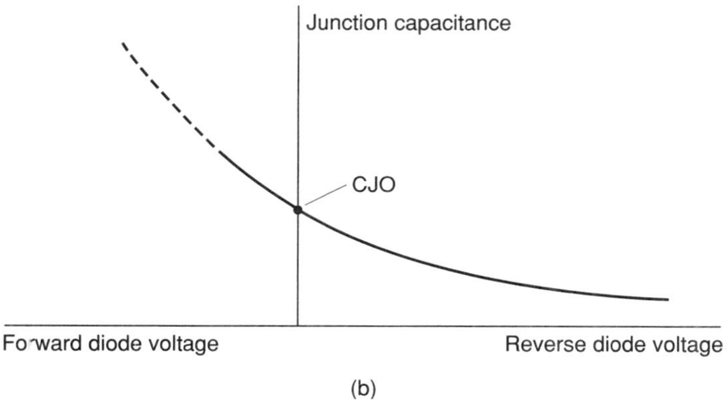

Figure 3.14(*b*) Determining the CJO model parameter.

and V_d is the diode voltage. If the value of VJ used for the x-axis variable is equal to the value of VJ needed for the model, the plot curve will be a straight line. If the value of VJ is too large or too small, the plot will be curved. This effect is shown in Fig. 3.15(*b*).

The only way to determine accurate values for M and VJ is to perform an iterative calculation. The procedure goes as follows:

1. Start with an initial estimate of VJ set to 1 V.
2. Plot the C-V data as shown in Fig. 3.15(*a*).

3*a*. If the graphed data forms a straight line, M will be equal to the slope of the line and VJ will be equal to the VJ used for the x-axis. Record these results and stop the procedure.

3*b*. If the graphed data forms a curved line, the value of VJ used to plot the graph must be increased or decreased. Increase or decrease VJ, then return to step 2.

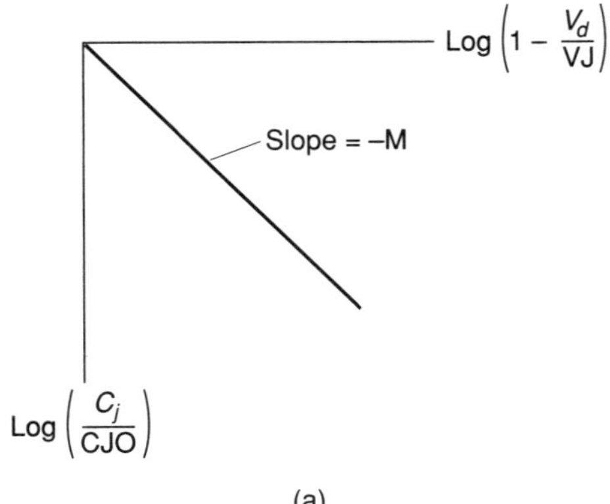

Figure 3.15(*a*) Graph to determine the model parameters M and VJ.

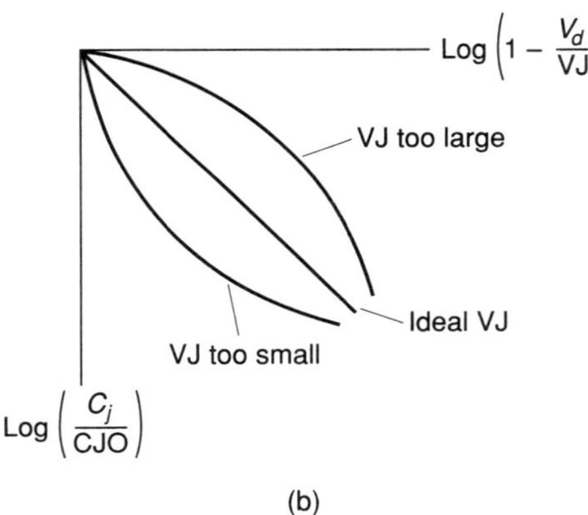

Figure 3.15(*b*) Effect of different values of VJ.

Continue graphing the data until you achieve a value of VJ which yields a linear plot.

While this graphing procedure can be extremely tedious to do by hand, it is easy to automate in a computer program. Just such a program has been created for you and has been included on the disk attached to this text. The program is called CJ_CALC. This program will ask you to enter the measured CJO value and several capacitance measurements for different diode voltages. The program follows the iterative procedure previously explained. The program saves the values of CJO, M, and VJ in a file named CJ_CALC.OUT, then generates two additional files, CJ_CALC.DAT and CJ_CALC.MAC. The former is a raw data file for the RGRAPH post processor (included on the disk). The latter is a command file for RGRAPH. After the program has found appropriate values for M and VJ, the user can call RGRAPH and display both the measured and simulated junction capacitance values. A typical RGRAPH plot from this program is shown in Fig. 3.16. Using the CJ_CALC program can greatly ease the procedure of extracting the capacitance parameters.

While it is beyond the scope of this chapter to give a detailed explanation of how to use CJ_CALC, the program is explained in detail in App. B. Since the program can be used in either an interactive mode or as a batch procedure, examples of both usages are given.

As previously stated, the capacitance coefficient model parameter FC will be set to 0.5 for all our diode models. Values much lower than this can degrade the model's accuracy and values much higher than this can lead to convergence problems.

To determine the TT model parameter from real devices can be extremely difficult without the aid of some very expensive measurement equipment and very special test setups. The reason for this is that the transit time for a diode can be on the order of nanoseconds. But we can obtain a good estimate of the transit time model parameter by using information from the diode data sheet. The TT model parameter can be estimated from the *reverse-recovery time* of the diode (T_{rr}). A plot of the reverse-recovery time vs. a ratio of forward-to-reverse current (I_r/I_f) is usually shown on the data sheet of switching diodes. An example of this is found in Fig. 3.18(*c*). Equation 3.15 and the reverse-recovery time can be used to estimate TT.

$$\text{TT} = 1.44 * T_{rr} \ (\text{at } I_r/I_f = 1) \tag{3.15}$$

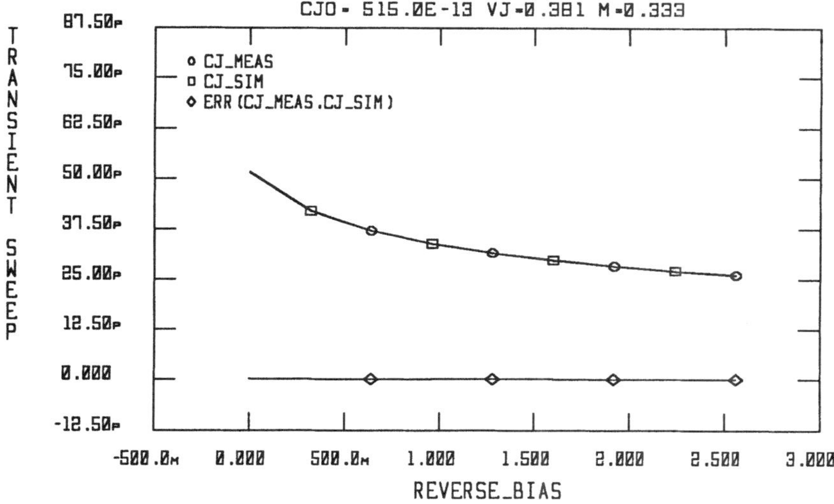

Figure 3.16 RGRAPH output from CJ_CALC program.

Temperature parameters

The temperature dependence of the diode is controlled by the parameters XTI and EG. The temperature coefficient XTI has a default value of 3, and the default value of EG is 1.11 eV (for silicon diodes). In contrast to the SPICE resistors where the user must set the temperature coefficients to a nonzero value before any temperature dependence is observed, the temperature coefficients of the diode are set by default to a nonzero value and the diode will exhibit temperature dependence automatically.

XTI is the diode temperature coefficient. The default value of 3 was chosen for silicon junction diodes. For Schottky barrier diodes, XTI should be set to 2. Extracting a new value is usually not necessary; however, sometimes, the value of XTI should be increased or decreased slightly to match a specific diode. Figure 3.17(*a*) and (*b*) illustrates how the XTI parameter influences the temperature dependence on the diode current.

EG is the energy band-gap voltage of the diode and is set according to the type of semiconductor being used. Most semiconductor physics books offer listings of the energy gap for different semiconductor materials. An abbreviated list of band-gaps for the most common semiconductors may be found in Massobrio.[3]

Diode Model Extraction Demonstration

To demonstrate the diode modeling procedures, several diodes and diode data sheets were collected. From these, two were chosen to illustrate our modeling procedures. The first diode we will model will be a 1N4002 general-purpose rectifier diode, and, to demonstrate the

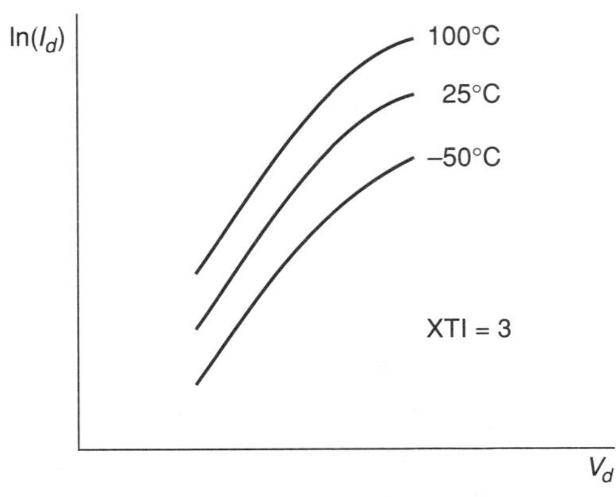

Figure 3.17(*a*) Temperature variations for XTI set to 3.

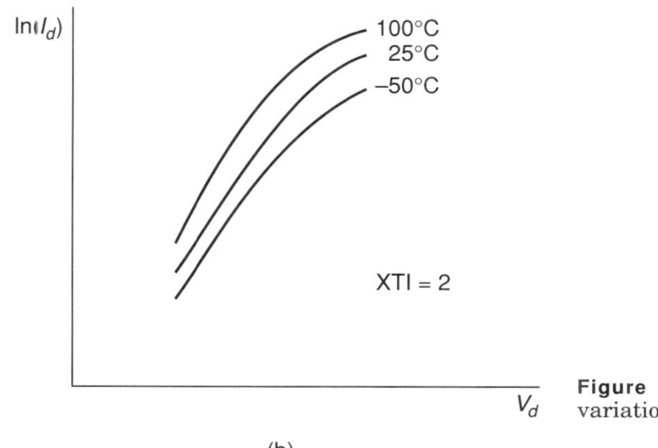

Figure 3.17(b) Temperature variations for XTI set to 2.

need for an improved zener model, a 1N752 5.7-V zener diode will be modeled in reverse operation.

To begin the extraction procedures, the 1N4002 diode data sheet was inspected. Three very useful pieces of information were immediately found. The first was a plot of the forward diode current vs. voltage, the second was a plot of the junction capacitance vs. diode voltage, and the third was a plot of the reverse recovery time. These plots are shown in Figs. 3.18(*a*), (*b*), and (*c*). From Fig. 3.18(*a*), IS, RS, and N can be obtained; from Fig. 3.18(*b*), CJO, M, and VJ may be determined; and from Fig. 3.18(*c*), TT can be estimated.

I-V parameters

From Fig. 3.18(*a*), 12 data points were observed. The values are shown in Table 3.4. From the diode current values, the natural log of the current can be calculated.

Once the voltage and current values were obtained, a plot of the natural log of the current vs. diode voltage was constructed. The plot is shown in Fig. 3.19. The curve appears to be linear for diode voltages between 0.58 and 0.9 V. To determine the IS and N parameters, the slope and intercept of this linear curve need to be determined. While you could do this manually from the plot, it is much faster and more accurate to use the linear regression program LINREG.

To help you extract model parameter values, a linear regression program has been developed and included on the book disk. The program is called LINREG.EXE. A batch-only version is also included, called LRB.BAT. The program can be used in an interactive mode, where the user is prompted for x and y values, or in a batch mode,

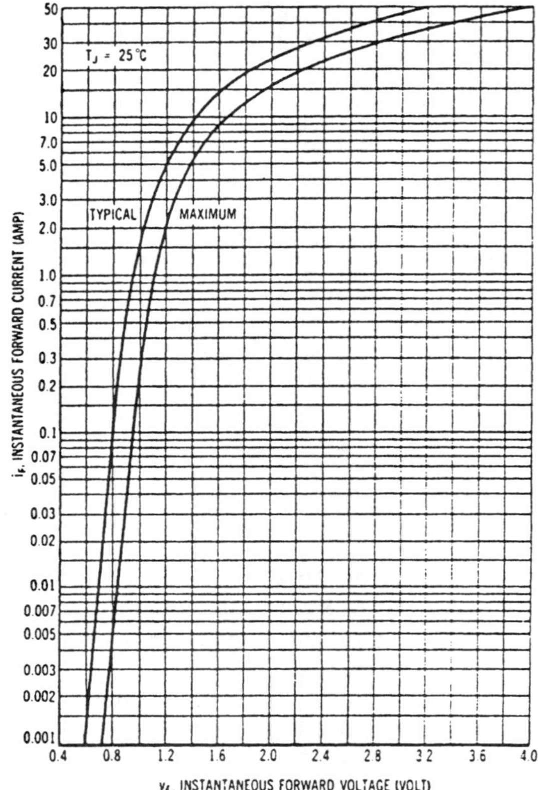

Figure 3.18(*a*) Forward voltage vs. diode current. (*Copyright Motorola; used by permission*)

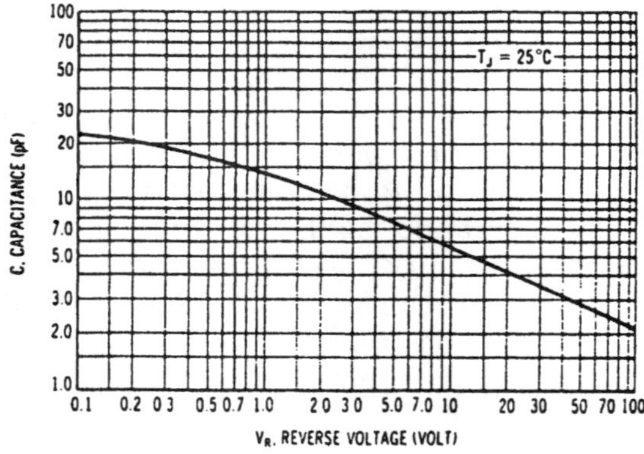

Figure 3.18(*b*) Reverse-biased voltage vs. junction capacitance. (*Copyright Motorola; used by permission*)

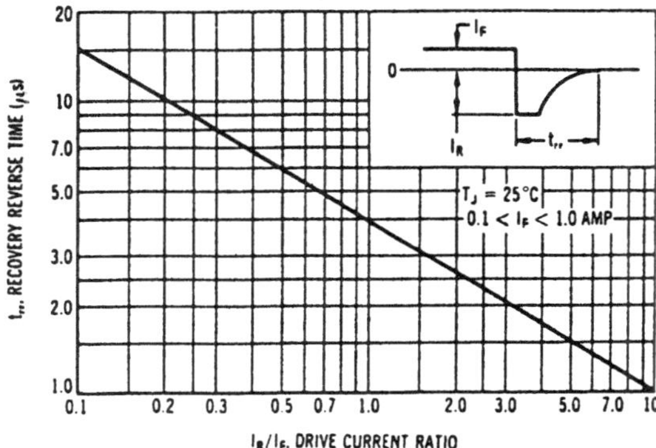

Figure 3.18(*c*) Reverse recovery time vs. drive current ratio. *(Copyright of Motorola; used by permission)*

TABLE 3.4 1N4002 I-V Data Taken from Data Sheet

V_d	I_d	$\ln(I_d)$
.58	.001	−6.9
.62	.002	−6.2
.66	.005	−5.3
.69	.010	−4.6
.72	.020	−3.9
.76	.050	−3.0
.80	.100	−2.3
.84	.200	−1.6
.90	1.0	0.0
.95	2.0	.69
1.2	5.0	1.6
1.4	10	2.3

where the values are stored in a text file. Using the program in both modes of operations is explained in App. C of this text.

Using the LINREG program found in the UTIL subdirectory of the book disk, both the slope and the *y*-axis intercept of Fig. 3.19 were determined. The slope of the linear approximation was 21.41. Using Eq. 3.13 gave an N value of 1.83. The intercept was −19.39, resulting in an IS of 3.8 nA. From the slope and intercept values, Table 3.2, and Eq. 3.14a–d, RS was calculated as follows:

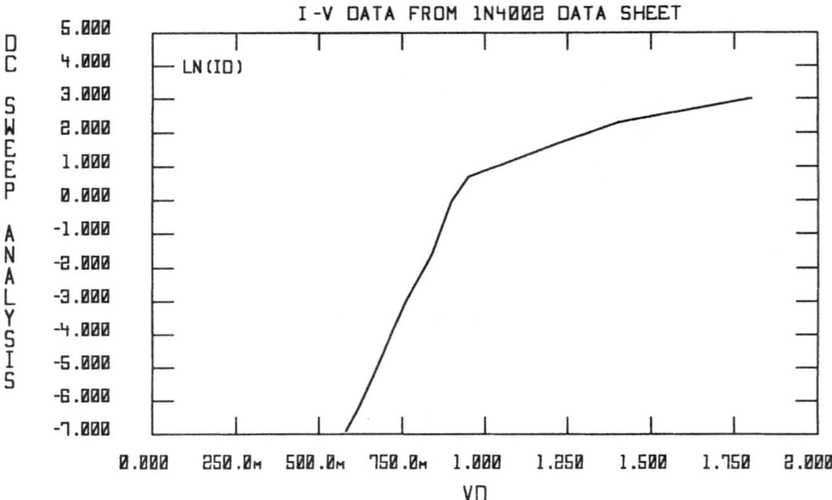

Figure 3.19 I-V data from 1N4002 data sheet.

$y_1 \;= \ln(10\text{ A}) = 2.3$	(from Eq. 3.14d)
$x_2 \;= 1.8\text{ V}$	(from Eq. 3.14c)
$x_1 \;= (y_1 - (y\text{-intercept}))/\text{slope} = 1.0$	(from Eq. 3.14b)
$\text{RS} = (x_2 - x_1)/10\text{ A} = 80\text{ m}\Omega$	(from Eq. 3.14a)

These model parameters were entered into a diode simulation. The diode voltage was swept from 0.58 to 3.2 V, while the diode current was measured. After the simulation was complete, the simulated values and the measured values were compared, as shown in Fig. 3.20(*a*) and (*b*). While these curves are similar, the rms error between the curves is a whopping 24.91 percent. See Fig. 3.20(*c*).

The reason for the error can be traced to reading data from the data sheet. The scale and grid on the graphs in Fig. 3.18(*a*) are too coarse to obtain an accurate reading (a reading accurate enough to obtain a 5 percent rms error model). *This example teaches us an important lesson in device modeling.* Reading I-V data from a data sheet will result in a less-than-ideal model. Reading data from a data sheet should be done only when actual I-V measurements cannot be performed!

Since our goal is to obtain a more accurate model (5 percent rms error), and because we have measurement equipment at our disposal, we will take measurements from several devices and use this I-V data and C-V data to make our model.

Armed with several 1N4002 diodes, I sat down in front of an HP4142 parametric tester and introduced myself. The HP4142 has four mea-

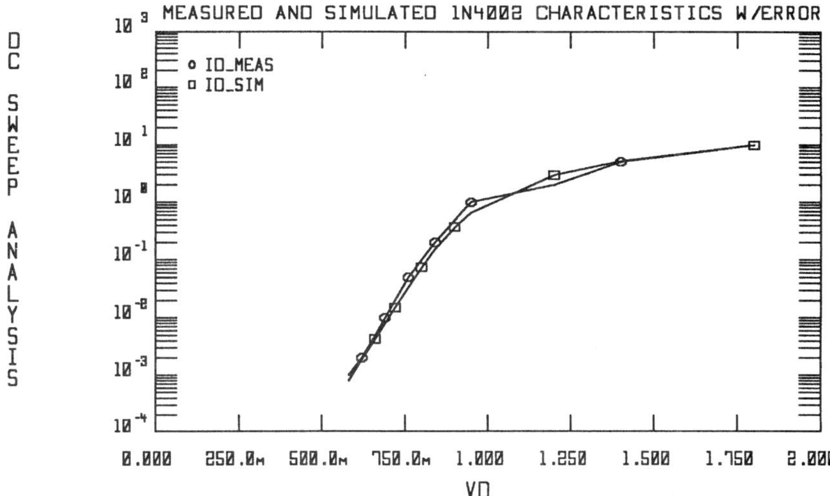

Figure 3.20(*a*) Measured and simulated 1N4002 characteristics.

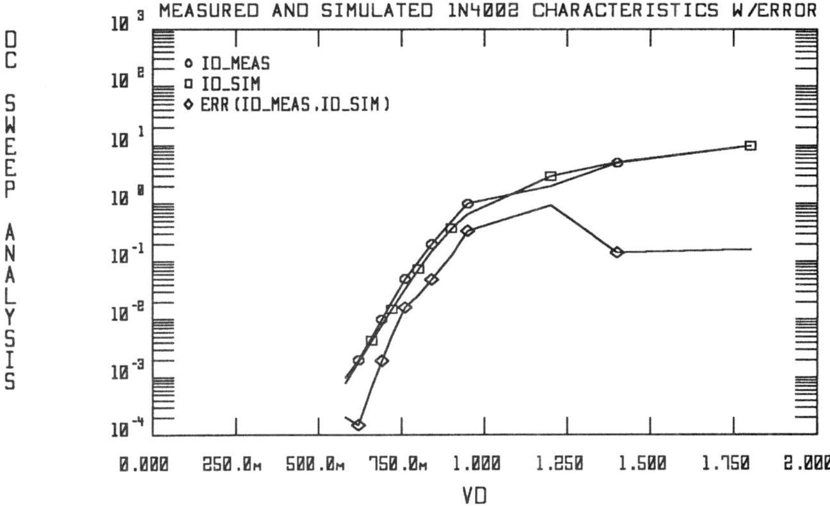

Figure 3.20(*b*) Measured and simulated 1N4002 characteristics with error.

surement units which allow you to force/measure current or voltage. This machine is ideal for making measurements for device modeling. The measurement setup for a diode is shown in Fig. 3.21.

A 1N4002 diode was tested and the I-V values recorded. The voltages and currents are shown in Table 3.5. Because of inherent limitations within the tester, diode voltages beyond 0.8 V could not be mea-

```
ERR('ID_MEAS','ID_SIM')    Percent Error


    10.0000E-04     79.6300E-05       20.4%
    20.0000E-04     18.5100E-04        7.5%
    50.0000E-04     42.9100E-04       14.2%
    10.0000E-03     80.3800E-04       19.6%
    20.0000E-03     14.9800E-03       25.1%
    50.0000E-03     33.7900E-03       32.4%
    10.0000E-02     73.6400E-03       26.4%
    20.0000E-02     15.0800E-02       24.6%
    50.0000E-02     37.1300E-02       25.7%
    10.0000E-01     66.0300E-02       34.0%
    20.0000E-01     29.3900E-01       46.9%
    50.0000E-01     51.4400E-01        2.9%
    10.0000E+00     98.3600E-01        1.6%

  RMS Error =    24.91%
```

Figure 3.20(*c*) Error between simulated and measured 1N4002 diode current.

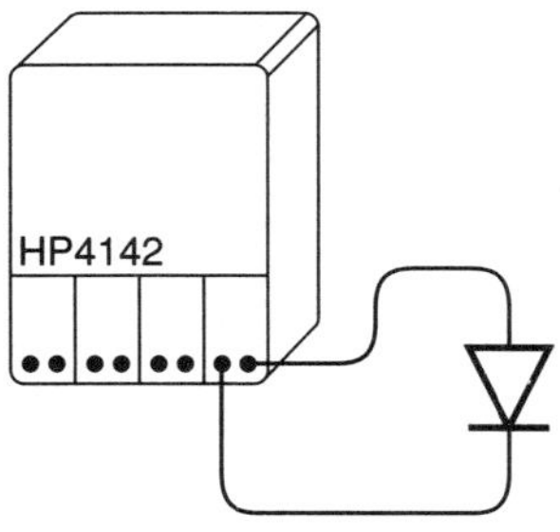

Figure 3.21 Test set-up for diode I-V measurements.

TABLE 3.5

V_d	I_d
2.000E-01	1.536E-08
2.500E-01	6.663E-08
3.000E-01	2.837E-07
3.500E-01	1.202E-06
4.000E-01	5.129E-06
4.500E-01	2.187E-05
5.000E-01	9.537E-05
5.500E-01	4.099E-04
6.000E-01	1.732E-03
6.500E-01	7.298E-03
7.000E-01	2.828E-02
7.500E-01	9.869E-02
8.000E-01	2.434E-01

sured. Notice that while an accurate model does not require a lot of data points, a good model requires *accurate* data.

As before, these values in Table 3.3 were entered into the LINREG program. A linear regression was performed on the data and found to produce a good fit for diode voltages between 0.2 and 0.5 V. The results of this regression are shown in Fig. 3.22, and a plot of the natural log of the diode current vs. diode voltage is shown in Fig. 3.23. From the regression results, the slope of the linear part of the curve was 29.06, and from Eq. 3.31, N was calculated as 1.35. The y-intercept was found to be −23.79 and IS was set to 46.5 pA. Finally, using the slope and intercept values, Table 3.3, and Eq. 3.14a–d, RS was calculated as follows:

$y_1 = \ln(0.2434\text{ A}) = -1.417$	(from Eq. 3.14d)
$x_2 = 0.8\text{ V}$	(from Eq. 3.14c)
$x_1 = (-1.417 - (y\text{-intercept}))/\text{slope} = 0.770$	(from Eq. 3.14b)
$\text{RS} = (x_2 - x_1)/0.2434\text{ A} = 123\text{ M}\Omega$	(from Eq. 3.14a)

These results were entered into the file 4002_IF.CIR (found in the CH3 subdirectory of the disk). The simulated and measured data are compared in Fig. 3.24(*a*) and (*b*). The rms error calculated between the simulated and measured data is only 1.24 percent, as shown in Fig. 3.24(*c*). While the error could probably be reduced by manually tweaking the parameters, the model has met our objective of less than 5 percent rms error, and we will let it be.

Under reverse-bias conditions, the diode current is constant. The value of the current is determined by −IS. In real diodes, the reverse-leakage current is not constant. Many designs can tolerate error in the reverse-bias region because the magnitude of the current is so small. So for now, in our 4002 diode model, we will model the diode current by IS; however, in the section labeled reverse-bias macromodel, the problem of a nonconstant reverse-leakage current will be addressed.

```
* * * * * * * * * * LINREG Results * * * * * * *
*
* For the equation Y = a + b*X
*
* a    =   -0.23795E+02
* b    =    0.29065E+01
* chi2 = 0.087
*
```

Figure 3.22 Linear regression output.

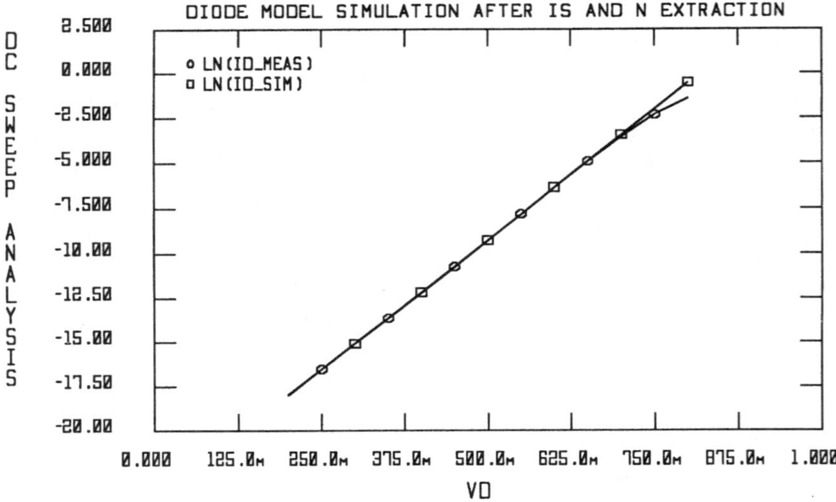

Figure 3.23 Diode simulation with IS and N parameters extracted from measured data.

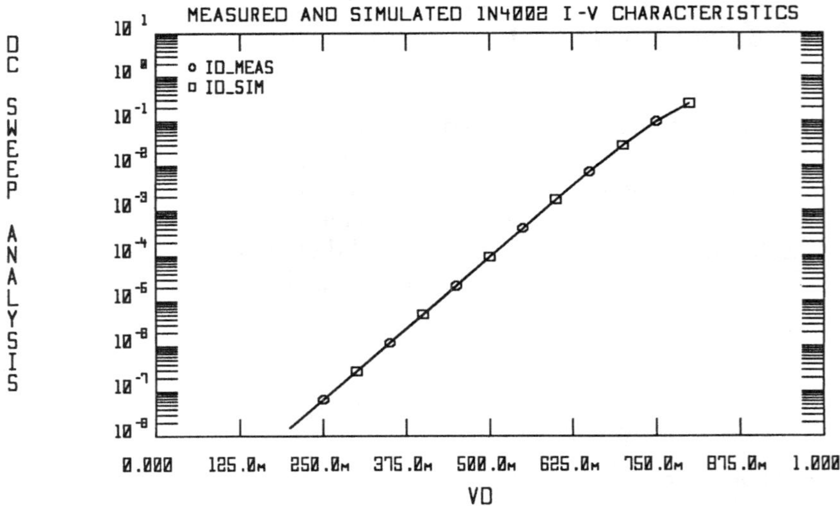

Figure 3.24(*a*) Diode model and simulation results after extracting IS, N, and RS.

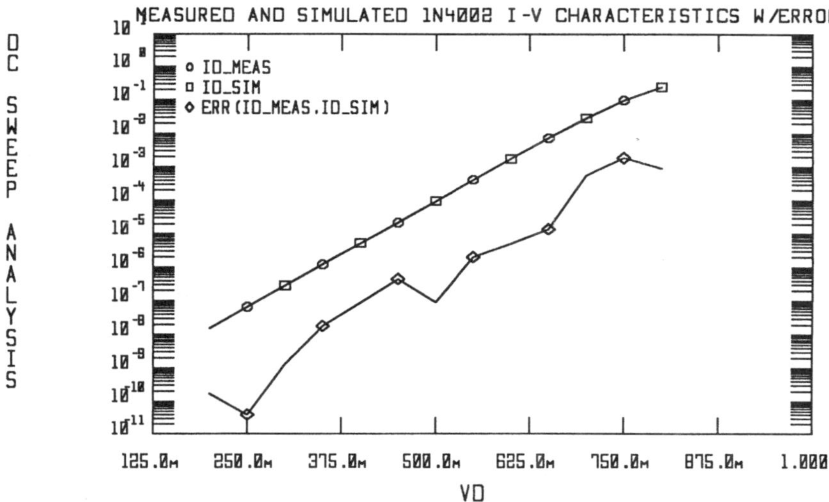

Figure 3.24(*b*) Diode model and simulation results including error between curves.

ERR('ID_MEAS','ID_SIM')		Percent Error
15.3600E-09	15.5300E-09	1.1%
66.6300E-09	66.5900E-09	0.1%
28.3700E-08	28.5000E-08	0.5%
12.0200E-07	12.2000E-07	1.5%
51.2900E-07	52.1800E-07	1.7%
21.8700E-06	22.3200E-06	2.1%
95.3700E-06	95.4600E-06	0.1%
41.0000E-05	40.8000E-05	0.5%
17.3200E-04	17.3700E-04	0.3%
72.9800E-04	72.8400E-04	0.2%
28.2800E-03	28.8500E-03	2.0%
98.6900E-03	96.8100E-03	1.9%
24.3400E-02	24.4300E-02	0.4%

RMS Error = 1.21%

Figure 3.24(*c*) Error report from forward I-V characteristics from 1N4002 diode.

Since the device is a rectifier diode, the parameters BV and IBV will be left at their default values of minus infinity and 1 mA. These parameters are used only for zener and avalanche breakdown.

Capacitance parameters

While Fig. 3.18(*b*) illustrates the reverse-bias junction capacitance curve, the curve is plotted on a logarithmic scale and does not include zero-biased junction capacitance (CJO). Since without CJO we cannot determine VJ or M, the graph is actually of little value. While the reader might try to estimate the value of CJO from the graph, the preferred remedy for this situation is to measure the junction capacitance of real devices and extract CJO, M, and VJ from the measured data.

A 1N4002 diode was placed in an HP4284 capacitance meter, as shown in Fig. 3.25. The junction capacitance was measured while the reverse-bias potential across the device was increased. Table 3.6 illustrates the measured capacitance-voltage values.

A graph of this data is shown in Fig. 3.26. In the discussion on calculating the reverse-bias junction capacitance values, a tedious graphing procedure was introduced to calculate these values. It was during this discussion that the programs CJ_CALC.EXE and CJB.BAT were introduced. CJ_CALC.EXE and CJB.BAT are discussed in detail in App. B. For those already familiar with the program or for those who want to skip right to the modeling part, the data in Table 3.6 was reformatted as shown in Table 3.7.

The batch version of CJ_CALC (known as CJB) was run on the data of Table 3.7. The values of CJO, M, and VJ calculated in the program are shown in Fig. 3.27. A plot of the measured and simulated data is shown in Fig. 3.28(*a*), and the corresponding error report is shown in Fig. 3.28(*b*). The CJB calculated model parameters yield a

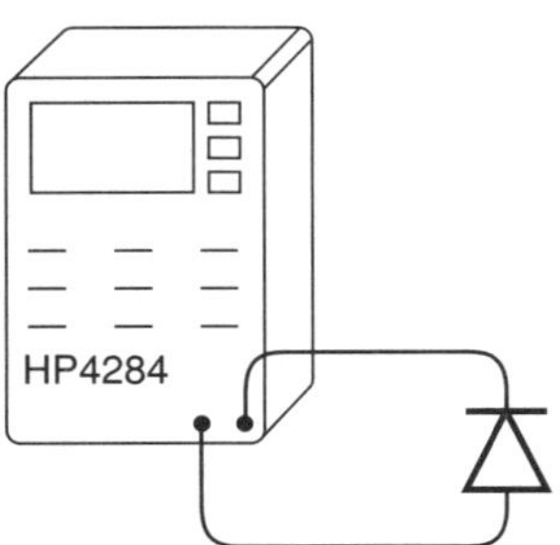

Figure 3.25 Test set-up to measure junction capacitance of 1N4002 diode.

TABLE 3.6 Measured Junction Capacitance vs. Voltage Bias

Reverse voltage	C_j
0	51.5E-12
.32	42.1E-12
.64	37.1E-12
.96	33.9E-12
1.28	31.6E-12
1.6	29.8E-12
1.92	28.3E-12
2.24	27.1E-12
2.56	26.1E-12

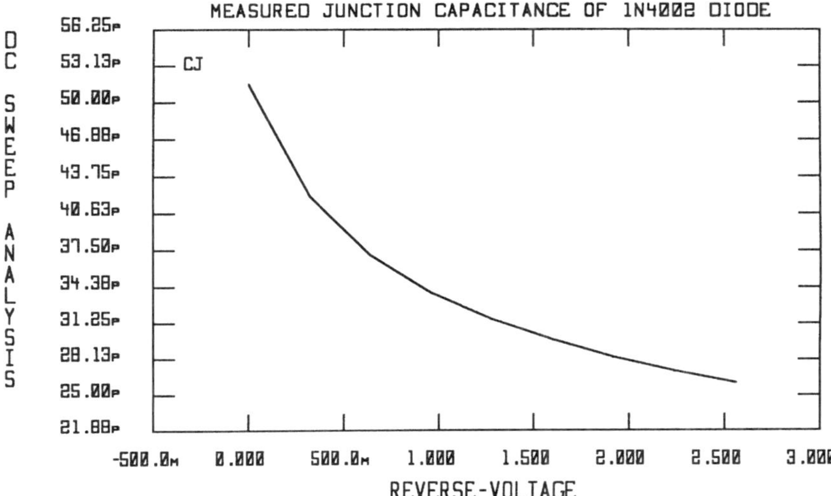

Figure 3.26 Measured junction capacitance of 1N4002 diode.

TABLE 3.7 Data Reformatted for the CJ_CALC and CJB Programs

19	! Number of samples
51.5E-12	! CJO
0	! Reverse bias voltage #1
51.5E-12	! Measured Cj
.32	! Reverse bias voltage #2
42.1E-12	! Measured Cj
.64	"
37.1E-12	
.96	
33.9E-12	
1.28	
31.6E-12	
1.6	
29.8E-12	
1.92	
28.3E-12	
2.24	
27.1E-12	
2.56	
26.1E-12	

very satisfying 0.16 percent rms error. Needless to say, we will leave these parameters alone.

With the reverse-bias junction capacitance model characterized, we can turn our attention to the dynamic capacitance TT. Inspecting Fig. 3.18(*c*), you will notice a plot of the reverse-recovery time (T_{rr}) vs. the

```
* * * * * * * * * * CJ_CALC Results * * * * * * * *
*
* cjo  =     515.000E-13
* vj   = 0.381
* m    = 0.333
*
```

Figure 3.27 Junction capacitance model parameters calculated from CJ_CALC.

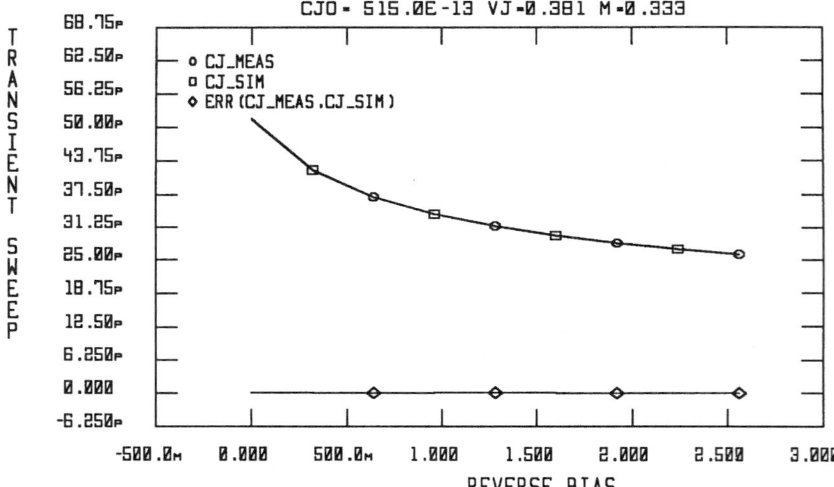

Figure 3.28(*a*) Diode simulation of junction capacitance and measured data with error.

```
ERR('CJ_MEAS','CJ_SIM')    Percent Error


    51.6000E-12      51.5000E-12          0.2%
    42.1000E-12      42.0274E-12          0.2%
    37.1000E-12      37.0775E-12          0.1%
    33.9000E-12      33.8579E-12          0.1%
    31.6000E-12      31.5280E-12          0.2%
    29.8000E-12      29.7308E-12          0.2%
    28.3000E-12      28.2841E-12          0.1%
    27.1000E-12      27.0835E-12          0.1%
    26.1000E-12      26.0639E-12          0.1%

  RMS Error =      0.16%
```

Figure 3.28(*b*) Error report for measured and simulated junction capacitance values of 1N4002 diode.

ratio of I_r to I_f. By locating the T_{rr} on the graph where I_r is equal to I_f, we can use Eq. 3.16 to calculate the transit time model parameter TT.

$$\text{TT} = 1.44{*}4\ \mu\text{S} = 5.77\ \mu\text{S} \tag{3.16}$$

We will use this value of transit time for our model.

Diode model summary

So far, we have calculated forward I-V and forward and reverse C-V characteristics to within our 5 percent rms error limit. The calculated model parameters are shown in Table 3.8. (*Note:* The parameters listed in Table 3.7 include only those whose values have changed from their default setting. The parameters AF, KF, XTI, EG, BV, and IBV remain at their default values.) For many designs, model development may stop here. While the reverse-leakage current may not meet our 5 percent rms error criterion because the current is so small, many designs can still be simulated with acceptable results.

Diode Macromodels

In some designs, the reverse-leakage current is crucial. For these types of circuits, a diode macromodel which accurately models the reverse current should be developed and employed. Two such diode macromodels are illustrated in Fig. 3.29(*a*) and (*b*).

The first model consists of a modeled diode DF, a leakage resistor RL, and two ideal diodes. The left side of Fig. 3.29(*a*) is used to model forward conduction of the real device. The extracted parameters RS, IS, N, CJO, M, and VJ should be assigned to the DF device. The leakage resistor RL can be used to model the reverse-leakage current of many large (high-current) devices. RL can be determined from the slope of the reverse-leakage current plot as shown in Fig. 3.30.

The ideal diodes are used as switches to block (1) the leakage current of the resistor when the diode macromodel is operated in the forward-bias region, and (2) the reverse-leakage current of DF when the

TABLE 3.8 1N4002 Diode Model Parameters

IS	= 46.5 pA
RS	= 123 MΩ
N	= 1.35
CJO	= 51.5 pF
M	= 0.333
VJ	= 0.381
FC	= 0.5
TT	= 5.77 μS

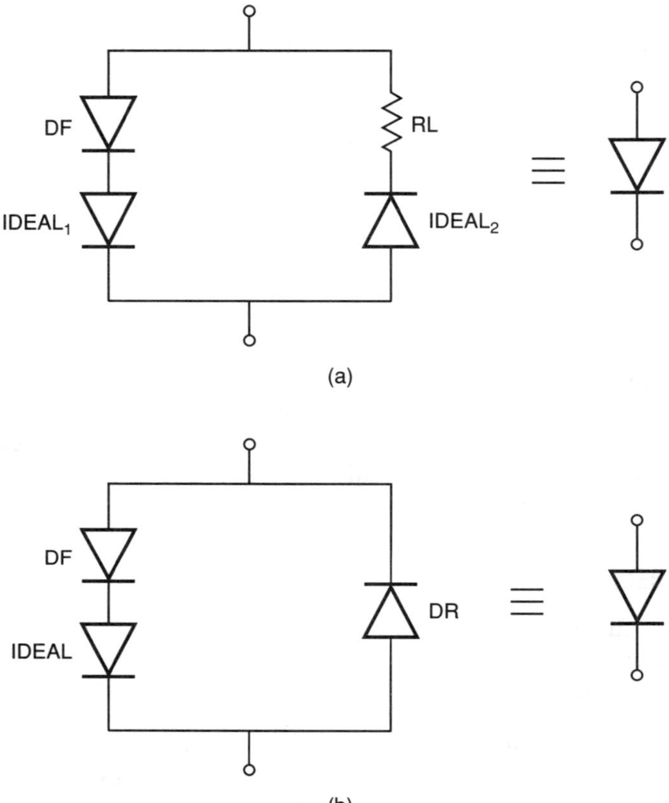

Figure 3.29 Macromodels for diode leakage current.

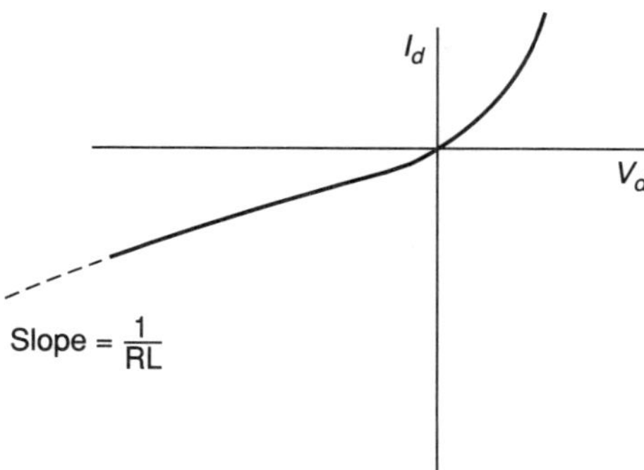

Figure 3.30 Determining the leakage resistance RL.

TABLE 3.9 Ideal Diode Model Parameters

N	= 0.001
RS	= 0.001
CJO	= 0.001 P
XTI	= 0*
EG	= 0

*Used only during temperature sweeps

diode is operated in reverse-bias. The ideal diode model should have the model parameters shown in Table 3.9.

Since diode leakage is actually characterized by an exponential relationship (although this relationship is very slight), the second macromodel can be used when an even more accurate reverse-leakage model is required. By measuring the reverse-leakage current vs. reverse-bias voltage characteristics, the IS and N values of DR can be extracted and used to model the reverse-leakage current.

To demonstrate the accuracy of the second macromodel, the reverse-bias I-V relationship of the 1N4002 diode was measured. The measured I-V values are shown in Table 3.10. These same values are plotted on a linear scale in Fig. 3.31(*a*) and on a natural log scale in Fig. 3.31(*b*).

TABLE 3.10 Reverse-Bias I-V Characteristics of 1N4002 Diode

V_d	$I_{(VID)}$
1.000E + 00	6.915E-11
1.500E + 00	1.213E-10
2.000E + 00	1.905E-10
2.500E + 00	2.822E-10
3.000E + 00	4.040E-10
3.500E + 00	5.656E-10
4.000E + 00	7.802E-10
4.500E + 00	1.065E-09
5.000E + 00	1.444E-09
5.500E + 00	1.947E-09
6.000E + 00	2.615E-09
6.500E + 00	3.502E-09
7.000E + 00	4.681E-09
7.500E + 00	6.248E-09
8.000E + 00	8.330E-09
8.500E + 00	1.110E-08
9.000E + 00	1.477E-08
9.500E + 00	1.966E-08
1.000E + 01	2.615E-08

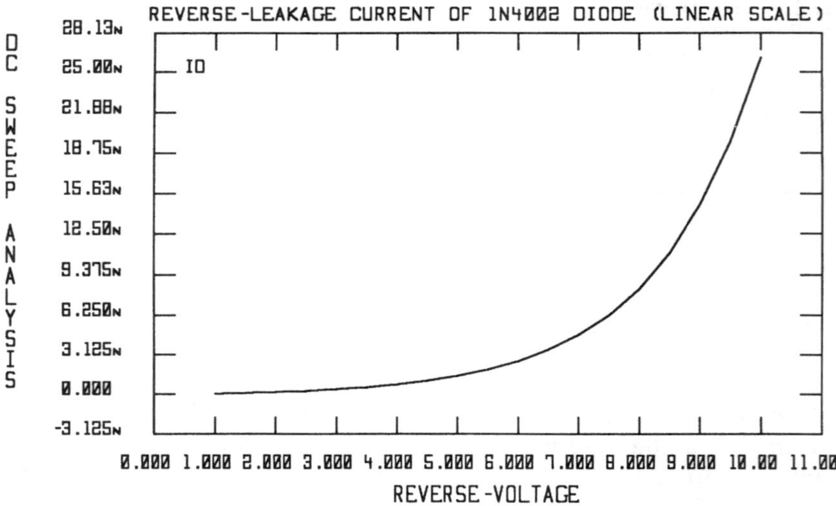

Figure 3.31(*a*) Reverse-leakage current of 1N4002 diode (linear scale).

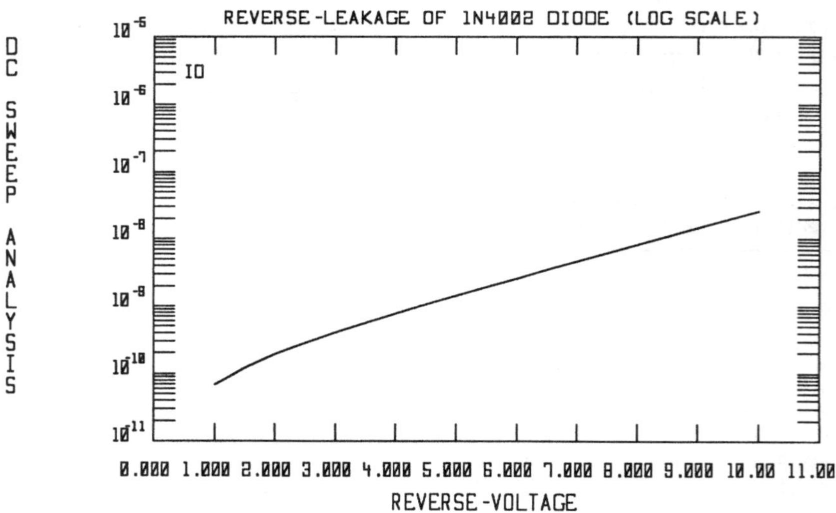

Figure 3.31(*b*) Reverse-leakage of 1N4002 diode (log scale).

While some diodes exhibit a linear characteristic in reverse-bias, this data is clearly exponential in nature and, therefore, fits a logarithmic curve very well. If the natural log of the current is plotted vs. the reverse-bias voltage, a linear plot will result. In fact, Fig. 3.31(*b*) is almost identical to Fig. 3.11. Following the same procedure used to extract the forward parameters IS and N, these parameters can be extracted and assigned to the diode DR. When the diode macromodel

TABLE 3.11 Reverse-Breakdown Characteristics of 1N752 Diode

Reverse voltage	I_d
0.2	1.44E-12
0.4	5.04E-12
0.6	8.76E-12
0.8	1.76E-11
1.0	3.38E-11
1.2	6.78E-11
1.4	1.32E-10
1.6	2.61E-10
1.8	5.31E-10
2.0	1.14E-09
2.2	2.62E-09
2.4	6.13E-09
2.6	1.42E-08
2.8	3.26E-08
3.0	7.18E-08
3.2	1.52E-07
3.4	3.11E-07
3.6	6.18E-07
3.8	1.19E-06
4.0	2.25E-06
4.2	4.21E-06
4.4	7.84E-06
4.6	1.45E-05
4.8	2.74E-05
5.0	5.31E-05
5.2	0.000108
5.4	0.000248
5.5	0.000413
5.6	0.000807
5.7	0.00250
5.8	0.0306
5.9	0.0798
6.0	0.136
6.1	0.193
6.2	0.24
6.4	0.32

is reverse-biased, diode DR is forward-biased and will predict the correct amount of reverse-leakage current for the macromodel.

Loading the I-V values of Table 3.9 into the LINREG program and performing a regression analysis yields a slope for this curve of 0.576 and a y-intercept of -23.16. From these values, the following model parameters can be found: IS = 87.3 pA, N = 68.01. Entering these values into the new diode macromodel, REVERSE.CIR, and simulating the reverse characteristics, yields the plot of Fig. 3.32. The error for this curve is 1.95 percent rms.

While both of these macromodels will generate a more accurate reverse-bias I-V characteristic than the stand-alone diode equations,

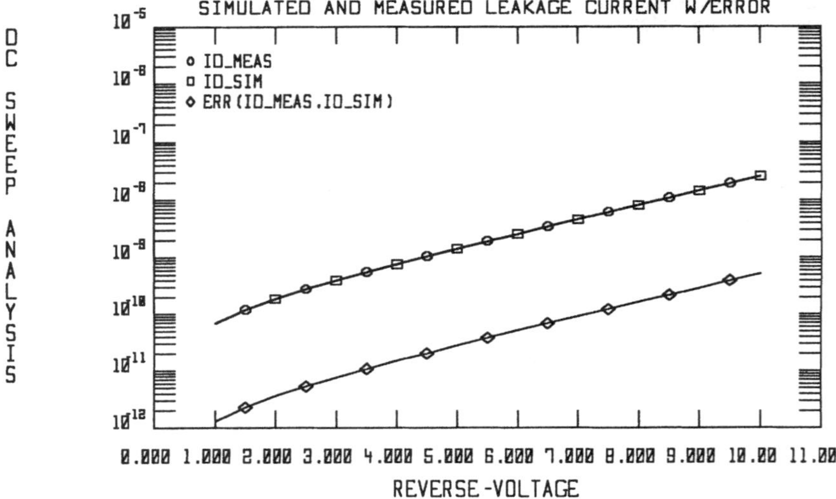

Figure 3.32 Simulated and measured leakage current with error.

the models do sacrifice the accuracy of the transit time (TT). Because of the separate conduction paths for forward- and reverse-bias current, the transit time will be incorrectly modeled. For this reason, you may want to select your model depending on your simulation application.

Zener diodes

To model zener and avalanche diodes, in addition to the forward characteristics discussed previously, the BV and IBV model parameters need to be extracted. While the authors of SPICE made the parameters easy to determine, to do so, they had to sacrifice accuracy. The parameter BV represents the voltage at which the diode enters breakdown. IBV represents the diode current at exactly the breakdown voltage. Extracting these parameters is as easy as selecting a breakdown voltage and measuring the current at exactly this point.

To demonstrate the extraction procedure, the reverse-breakdown characteristics of a 5.7-V 1N752 zener diode were measured on the HP4142. The recorded data is shown in Table 3.11. Using a spreadsheet program, the point of maximum curvature can be determined as shown in Fig. 3.33. From this data, we will choose BV to be 5.7 V, and from the table, IBV is 0.0025 A.

These values were entered into the ZENER1.CIR circuit file and simulated. The measured and simulated reverse-breakdown characteristics are illustrated in Fig. 3.34(*a*)–(*c*).

The BV and IBV model parameters, although easy to extract, fail to offer a way to tailor the shape of the I-V characteristics. The shape is

Reverse-bias Voltage	Reverse Id	ln(Id)	Percent Change in current from last point
0.2	1.44E-12	-27.2664	0
0.4	5.04E-12	-26.0136	-71.43%
0.6	8.76E-12	-25.4608	-42.47%
0.8	1.76E-11	-24.7631	-50.23%
1	3.38E-11	-24.1106	-47.93%
1.2	6.78E-11	-23.4145	-50.15%
1.4	1.32E-10	-22.7482	-48.64%
1.6	2.61E-10	-22.0665	-49.43%
1.8	5.31E-10	-21.3563	-50.85%
2	1.14E-09	-20.5922	-53.42%
2.2	2.62E-09	-19.7601	-56.49%
2.4	6.13E-09	-18.9101	-57.26%
2.6	1.42E-08	-18.07	-56.83%
2.8	3.26E-08	-17.239	-56.44%
3	7.18E-08	-16.4494	-54.60%
3.2	1.52E-07	-15.6994	-52.76%
3.4	3.11E-07	-14.9835	-51.13%
3.6	6.18E-07	-14.2968	-49.68%
3.8	1.19E-06	-13.6416	-48.07%
4	2.25E-06	-13.0046	-47.11%
4.2	4.21E-06	-12.378	-46.56%
4.4	7.84E-06	-11.7563	-46.30%
4.6	1.45E-05	-11.1414	-45.93%
4.8	2.74E-05	-10.505	-47.08%
5	5.31E-05	-9.84333	-48.40%
5.2	0.000108	-9.13338	-50.83%
5.4	0.000248	-8.30208	-56.45%
5.5	0.000413	-7.79134	-40.00%
5.6	0.000807	-7.12219	-48.79%
5.7	0.0025	-5.99146	-67.72% } Maximum curvature
5.8	0.0306	-3.48676	-91.83% }
5.9	0.0798	-2.52823	-61.65%
6	0.136	-1.9951	-41.32%
6.1	0.193	-1.64507	-29.53%
6.2	0.24	-1.42712	-19.58%
6.4	0.32	-1.13943	-25.00%

Figure 3.33 Determining the BV and IBV model parameters.

determined by the exponential of Eq. 3.4. The impact of this inflexibility is observed in Fig. 3.34. For most simulations, the amount of error in Fig. 3.34 is completely unacceptable.

To overcome this difficulty, a new diode macromodel can be developed. The new model starts with the macromodel of Fig. 3.29 and includes an independent voltage source and a resistor to represent the breakdown voltage. The new macromodel is shown in Fig. 3.35.

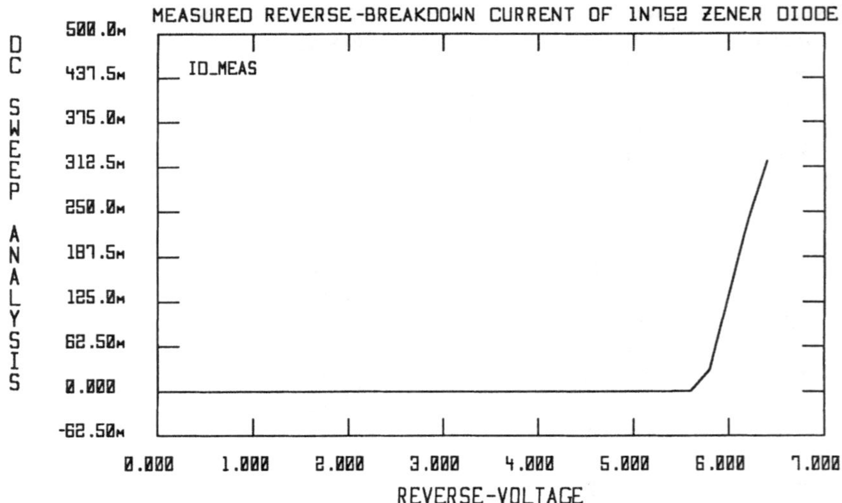

Figure 3.34(*a*) Measured and simulated zener breakdown with the BV and IBV parameters.

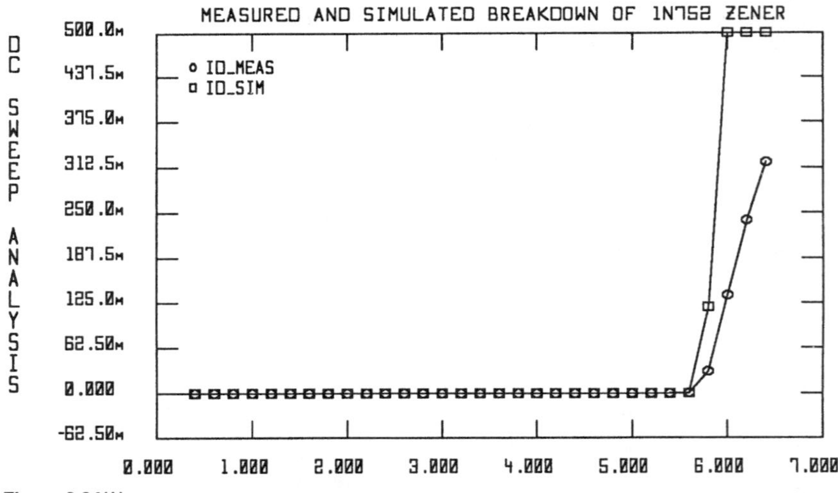

Figure 3.34(*b*)

Under reverse-bias conditions, the diode DR begins to conduct. As the reverse-biased potential exceeds the breakdown voltage, V_{BRK} the ideal diode $DIDEAL_2$ closes the circuit to the zener resistor, and the resistor begins to conduct current.

Again looking at the reverse-biased 1N752 zener data, the I-V data from 0.2 to 5.7 V is exponential in nature, but from 5.7 to 6.4 V, the data becomes almost linear in character. This means the I-V charac-

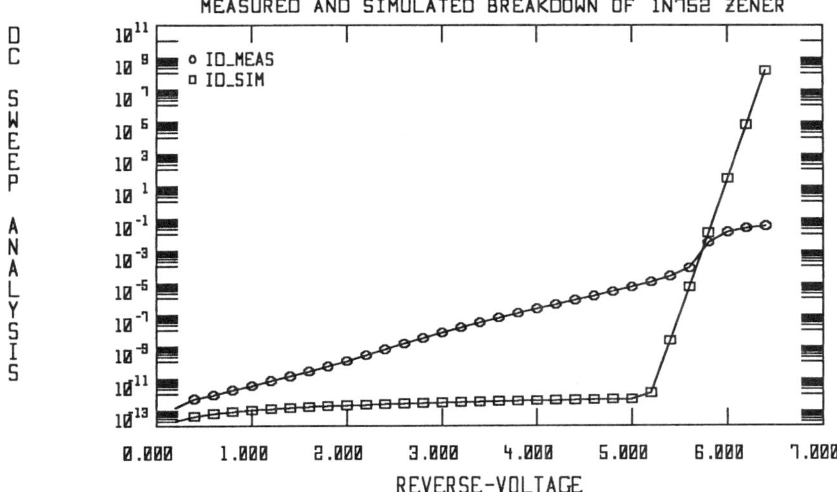

Figure 3.34(*c*) Measured and simulated zener breakdown plotted on a logarithmic scale.

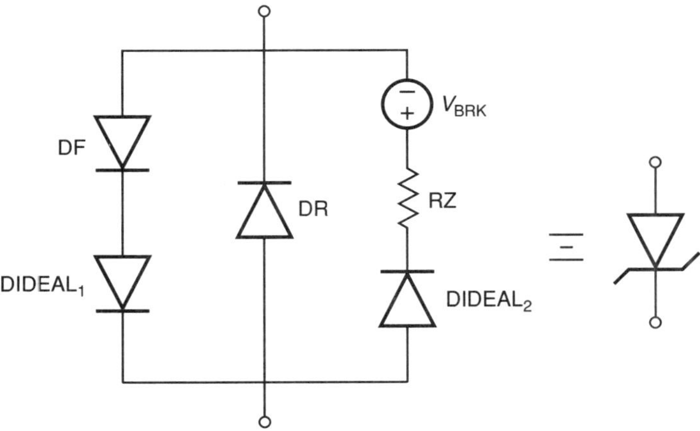

Figure 3.35 Zener diode macromodel.

teristics from 0.2 to 5.7 V can be described with an exponential and the characteristics from 5.7 to 6.4 V can be described with a simple resistor. Using linear regression on the natural log of the reverse-bias current from 0.8 to 5.4 V, we calculate IS = 1.0 pA, N = 10.85. Using linear regression on the reverse-biased current from 5.7 to 6.4 V yields RZ = 2.1 Ω.

These new values were entered into a circuit named ZENER2.CIR and simulated. Figures 3.36(*a*) and (*b*) illustrate that the zener macromodel yields a much more acceptable level of accuracy. The

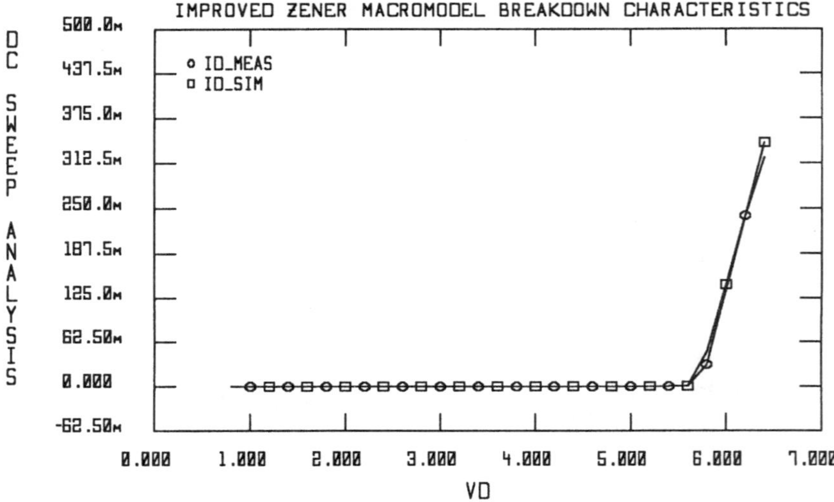

Figure 3.36 Measured and simulated zener breakdown of zener macromodel.

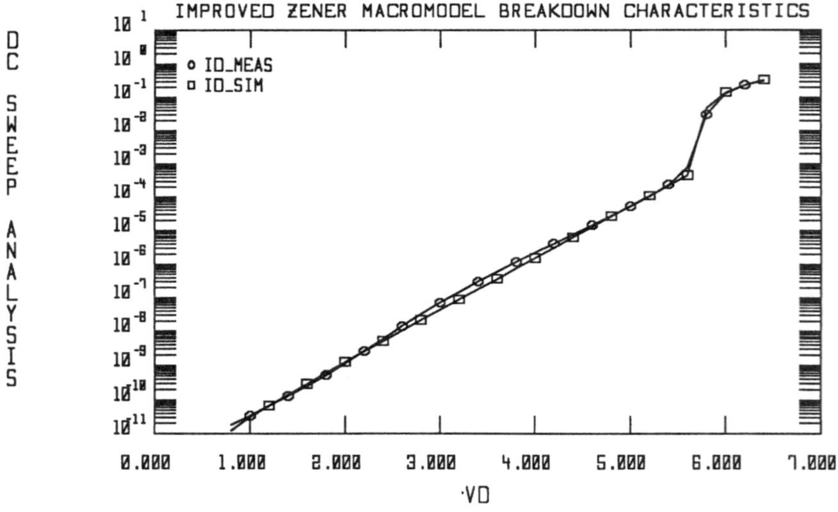

Figure 3.36 *(Continued)*

models used for this example, ZENER2.MOD and ZENER2.CIR, are stored in the CH3 subdirectory of your book disk.

The zener macromodel presented here greatly increases the accuracy of the reverse-leakage and reverse-biased breakdown current. The models are easy to construct, and the parameters of the model are easy to obtain. While the macromodels do sacrifice the accuracy of the diode model's transit time, many zener and avalanche diodes are operated only in the reverse region and never see forward-bias or switching conditions. For these devices, the zener macromodels are a valid option. If a zener diode is used in both forward- and reverse-bias operation, one faces the decision of whether inaccuracy in the switching region or inaccuracy in the reverse-leakage, reverse-breakdown current is more tolerable. The model chosen will depend on this decision. The zener macromodels presented here significantly increase the accuracy of the standard diode model in SPICE.

Temperature-dependent breakdown

For many zener and avalanche diodes, the breakdown voltage is a function of the temperature of the device. In addition, while some devices may have a positive temperature coefficient, others may have a negative coefficient. These effects can easily be added to our zener macromodel, as shown in Fig. 3.37.

The model in Fig. 3.37 is identical to the model in Fig. 3.35 except the independent voltage source used to represent V_{BRK} has been replaced with a dependent voltage source E_{BRK}. The voltage of this source is determined by the resistor RT, the temperature coefficient of RT, and the temperature of the circuit. For a positive breakdown coefficient, the temperature coefficient of RT must be greater than zero. For a negative coefficient, the temperature coefficient of RT must be less than zero. To adjust the breakdown voltage of the device, the resistance value of RT is adjusted. This model is extremely flexible and easy to use. The characteristics of the temperature-dependent model are illustrated in Fig. 3.38.

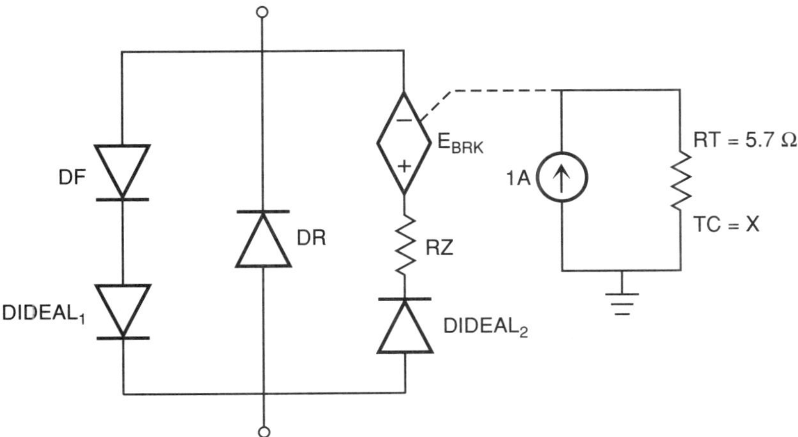

Figure 3.37 Temperature-dependent breakdown macromodel.

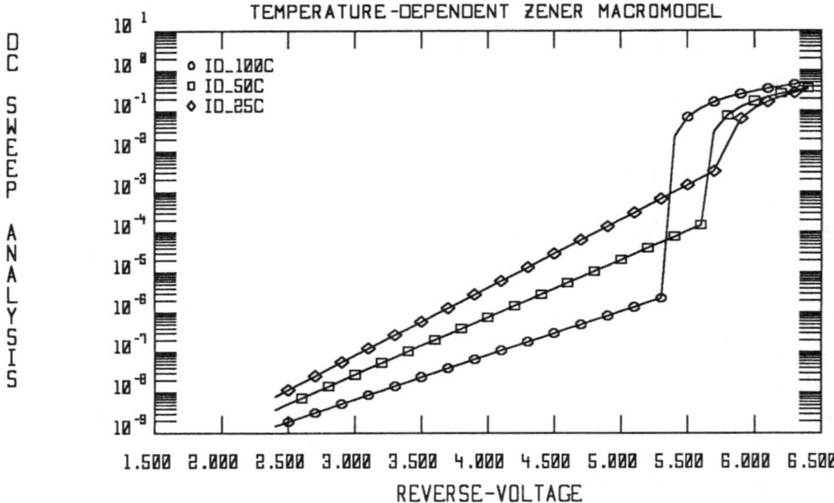

Figure 3.38 Temperature-dependent Zener output characteristics.

Diode Summary

This chapter has focused on learning to extract diode model parameters from measured data. In this section we also learned that, while it is easy to read the I-V and C-V data presented on device data sheets, often these readings are not accurate enough to meet our goal of 5 percent rms accuracy. While values read from a data sheet may be used to make a first-order model, an accurate model—one with 5 percent rms error or less—will probably require measured data.

The foundation of the diode model, the DC parameters, are extracted from simple linear approximations of the I-V characteristics of the device. To help readers develop the DC parameters, the LINREG program was developed and is included on the book disk. The nonlinear C-V parameters are slightly more difficult to extract than their DC cousins. The capacitive parameters require an iterative solution technique. A graphing procedure was introduced to calculate the junction capacitance terms. Again, to aid the reader in developing the capacitance parameters, the CJ_CALC program was developed and is included on the book disk. Because of the difficulty in measuring TT accurately, a simple estimation process was introduced. To help improve the accuracy of the model for reverse-bias leakage current, a macromodel was introduced. For zener diodes and other breakdown effects, the BV and IBV parameters can be used, but the accuracy of the reverse-breakdown current will be poor. More accurate zener simulations require a zener macromodel. The zener macromodels are easy to construct and significantly improve the accuracy of the simulation.

While not all users will need the level of accuracy afforded by the various macromodels, it is good to know they are available when needed.

The modeling procedures and macromodels presented in this chapter demonstrate that accurate modeling is neither difficult nor does it take much time. Throughout this chapter, tables of I-V and C-V data have been presented so the reader can experiment with the procedures and techniques documented here.

References

1. Ron Kielkowski, *Inside SPICE: Overcoming the Obstacles of Circuit Simulation,* McGraw-Hill, 1994.
2. Lawrence Nagel, *SPICE User's Guide,* Memorandum No. ERL-M520, May 1975, University of California, Berkeley.
3. Giuseppe Massobrio and Paolo Antognetti, *Semiconductor Device Modeling with SPICE,* McGraw-Hill, 1988, 1993.

Chapter

4

Modeling the Bipolar Junction Transistor

Introduction

In this chapter, you will learn to model the bipolar junction transistor (bjt). The bjt used in SPICE is modeled with the equivalent circuit shown in Fig. 4.1(*a*), and as such, many of the model parameters and extraction techniques you learned in the previous section will be applied to the base-emitter (b-e) and base-collector (b-c) junctions. For these junctions, you will be extracting parameters like the reverse-leakage current (IS), the emission coefficient (N) and the series resistance (RS). You will also extract the zero-bias capacitance parameter (CJO), the grading coefficient (M) and the built-in potential (VJ). Each of these parameters must be determined for the b-e and the b-c junctions. For integrated circuits, these same capacitance parameters must be determined for the collector-substrate (c-s) junctions [the equivalent circuit for an integrated circuit bipolar transistor is shown in Fig. 4.1(*b*)]. There are 40 bipolar model parameters, and, of those 40, 15 are basic diode parameters. This is why understanding the diode parameters is essential to understanding the bipolar transistor model.

If you have already read Chap. 3 of this text, which dealt with modeling junction diodes, you are now ready to model a bjt. If you skipped Chap. 3 and turned directly to this section, you should first read the section on modeling the diode. The information learned about the diode parameters and the modeling and extraction techniques covered in the diode chapter will be used in this chapter and many others. The diode chapter also introduced the programs CJ_CALC and LINREG. We will make extensive use of these programs throughout

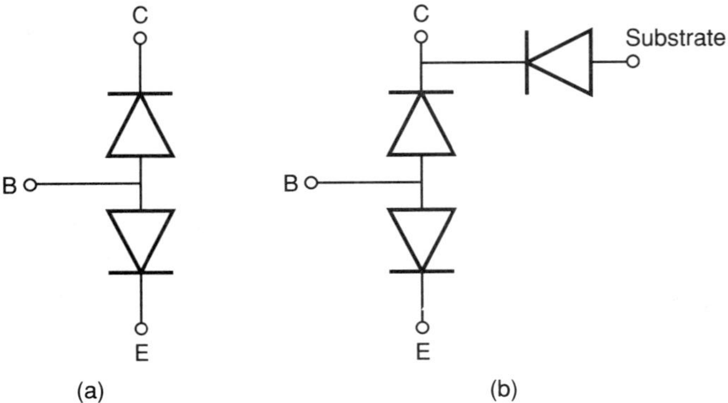

Figure 4.1(*a*) SPICE bipolar equivalent circuit for discrete transistors. **(*b*)** SPICE bipolar equivalent circuit for integrate circuit transistors.

this chapter, too.

Throughout this chapter, various circuits and models will be described. All of the circuit and models are found in the CH4 subdirectory of the disk bound to this text. The utility programs CJ_CALC, CJB, LINREG, and LRB are found in the UTIL subdirectory of the disk. The disk must be installed on your hard disk and uncompressed before you can access these files.

As we have throughout this text, we will focus on building *practical* bjt models in this chapter. Some of the parameters of the bipolar transistor are outside the bounds of what is practical. These parameters include RBM, IRB, XTF, VTF, ITF, and PTF. These parameters are buried so far into the bipolar equations that there is simply no good method of making simple measurements to obtain these values. For designers who require the utmost in accuracy, these parameters can be determined with an optimizer. (An optimizer extracts model parameters from measured I-V and C-V data.) But in most cases, a practical model of the bipolar transistor can be made without them.

We will begin by looking at the standard SPICE bjt model parameters, defining and describing each in detail and explaining how each one is extracted. Following the explanation, a real bipolar transistor will be modeled. The SPICE bipolar transistor model is well suited to many discrete and integrated-circuit transistors.

What you will find notably missing in this chapter is a detailed examination of the equations used to model the bjt. While some may criticize this decision, the author has chosen to eliminate these from this chapter for several reasons: (1) the bipolar equations are quite complicated and involve a nontrivial amount of device physics to understand; (2) the equations have been described in detail in Refs. 1, 2, and 4, so to reprint them here would be redundant; and (3) knowledge of the equations is not required to understand how to extract model parameters. While knowledge of the equations is required to develop the extraction procedures, once a suitable extraction technique is adopted, knowledge of the derivation is not needed to extract a parameter—all you need then is knowledge of how to perform the extraction. In keeping with our philosophy of practical device modeling, this text will contain only the bipolar equations necessary to understand the extraction procedures. Readers interested in the complete set of bipolar equations are referred to Refs. 1, 2, and 4.

Modeling Approach

In accordance with our philosophy of practical device modeling, typical bipolar transistor data sheets will be inspected for information which might be useful in developing device models. What cannot be obtained from the data sheet will be taken from measured I-V and C-V characteristics. After the initial set of model parameters has been chosen, simulations will be run on the model to optimize the final parameter values and to estimate the accuracy of our model.

Bipolar Model Parameters

Again, an introduction to the model parameters is necessary. Table 4.1 lists the 40 bipolar transistor model parameters.

Also shown in Table 4.1, along with the name, is the default value of each parameter. Every model parameter has a default value assigned inside the program. The parameter values found on the .MODEL statements simply override the default values assigned within the program.

The first 18 parameters, IS, BF, *BR*, NF, *NR*, ISE, *ISC*, IKF, *IKR*, NE, *NC*, VAF, *VAR*, RC, RE, RB, RBM, and IRB, describe the DC I-V characteristics of the bjt. Of these 18, the 6 parameters shown in italic represent the "reverse" parameters of the bjt. From a *practical* perspective, these parameters are used only when the transistor is operated in the reverse region of operation. These reverse parameters are equivalent to their corresponding forward parameters only for reverse voltages. The forward and reverse regions of operation are shown in Fig. 4.2.

The next set of 17 parameters, CJE, MJE, VJE, CJC, MJC, VJC, CJS, MJS, VJS, XCJC, FC, TF, TR, XTF, VTF, ITF, and PTF, describe the capacitive characteristics of the model. Of these, CJx, MJx, and VJx describe the reverse-bias junction capacitance for the base-emitter (x = E), base-collector (x = C), and, for integrated-circuit bjts, the collector-substrate (x = S) junctions. TF, TR, and the XTF parameters are used to model the reverse-recovery time of the transistor.

TABLE 4.1 bjt Model Parameters

Name	Parameter	Units	Default
	DC Parameters		
IS	Saturation current	V	1e-16
BF	Forward beta	—	100
BR	Reverse beta	—	1
NF	Forward emission coefficient	—	1
NR	Reverse emission coefficient	—	1
ISE	b-e saturation current	A	0
ISC	b-c saturation current	A	0
IKF	Forward high-current beta roll-off	A	∞
IKR	Reverse high-current beta roll-off	A	∞
NE	b-e emission coefficient	—	1.5
NC	b-c emission coefficient	—	2.0
VAF	Forward early voltage	V	∞
VAR	Reverse early voltage	V	∞
RC	Collector resistance	Ω	0
RE	Emitter resistance	Ω	0
RB	Base resistance	Ω	0
RBM	Minimum base resistance	Ω	RB
IRB	Current at half minimum base resistance	A	∞
	Capacitance Parameters		
CJE	b-e zero-bias junction cap.	F	0
MJE	b-e grading coefficient	—	0.33
VJE	b-e built-in potential	V	0.75
CJC	b-c zero-bias junction cap.	F	0
MJC	b-c grading coefficient	—	0.33
VJC	b-c built-in potential	V	0.75
CJS	c-s zero-bias junction cap.	F	0
MJS	c-s grading coefficient	—	0.33
VJS	c-s built-in potential	V	0.75
XCJC	Fraction of b-c capacitance at base node	—	1
FC	Coefficient for capacitance	—	.5
TF	Forward (b-e) transit time	S	0
TR	Reverse (b-c) transit time	S	0
XTF	Coefficient for bias dependence of TF	—	0
VTF	Voltage dependence of Vbc on TF	V	∞
ITF	High-current parameter for TF	A	0
PTF	Excess phase at f = 1/2pi*TF	Degrees	0
	Temperature Parameters		
XTB	Beta temperature coefficient	—	0
XTI	Saturation current temperature coefficient	—	3
EG	Energy-band gap for temperature effect	eV	1.11
	Noise Parameters		
KF	Flicker-noise coefficient	—	0.0
AF	Flicker-noise exponent	—	1.0

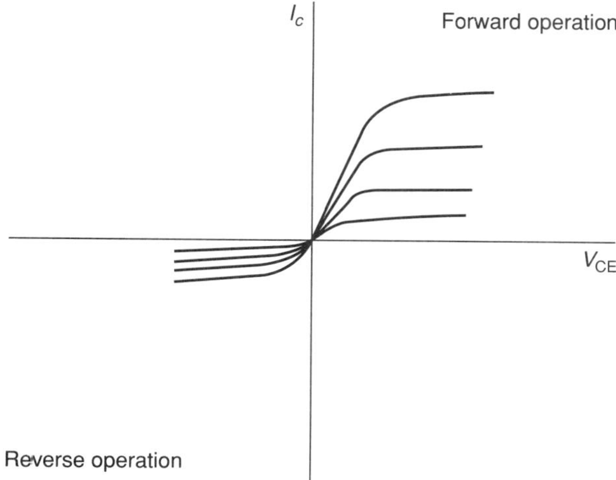

Figure 4.2 Forward and reverse regions of operation.

The parameters EG, XTI, and XTB model the temperature-dependent changes in bjt current and capacitance, while KF and AF model the flicker-noise of the device (KF and AF are used only during an AC noise analysis).

These 40 model parameters completely characterize the bipolar transistor in SPICE. The parameters can be used to model both discrete and integrated-circuit transistors. A practical bjt model can be built for almost any transistor from these 40 parameters.

As with the diode, many accurate device models can be built from something less than all 40 parameters. For example, the six reverse parameters, BR, NR, ISC, NC, IKR, and VAR, are used only when the transistor is used in reverse operation. In normal forward operation, these parameters have very little influence on the model. For transistors operated in the normal mode, these reverse parameters may be left at their default values. EG, XTI, and XTB are used only during SPICE temperature sweeps or analyses at different temperatures. CJS, MJS, and VJS are required only for integrated-circuit transistors which have a substrate contact. Thus, many bipolar transistor models can be built from only a subset of the total 40 model parameters.

The DC bipolar parameters

The most basic and most important parameters of the bipolar model are IS, NF, and BF. These three parameters form the foundation of the entire model. To understand these parameters, we need to look at the collector and base current equations of the transistor, shown in

Eqs. 4.1 and 4.2.

$$I_c = \text{IS} * \left[\exp\left(\frac{q * V_{\text{be}}}{\text{NF} * k * t}\right) - 1\right] \tag{4.1}$$

$$I_b = \frac{I_c}{\text{BF}} \tag{4.2}$$

As might be expected from Fig. 4.1(*a*) or (*b*), the collector current is basically an exponential function of the base-emitter voltage, and Eq. 4.1 is nearly identical to the diode current expression of Chap. 3. BF (otherwise known as h_{FE} to transistor designers) defines the ratio of the collector and base currents.

To extract these parameters, we need the DC I-V characteristics of the transistor. While there are several I-V relationships we can define from the base, emitter, and collector terminals of the transistor, the most common is the family of curves shown in Fig. 4.3(*a*). The family of curves can be obtained from the measurement setup shown in Fig. 4.3(*b*). Unfortunately, the family of curves does not describe the complete I-V relationship of the transistor. A better measurement for model parameter extraction is known as the Gummel test, which results in the Gummel plot.[3] The Gummel plot is shown in Fig. 4.4(*a*) and the test setup is shown in Fig. 4.4(*b*). The model parameters IS, NF, and BF are easily obtained from the Gummel plot.

The Gummel plot illustrates the base and collector current as the base-emitter voltage (V_{be}) changes (with the base-collector voltage (V_{bc}) at zero). The natural log of the base and collector currents are plotted on the *y*-axis with the base-emitter voltage plotted on the *x*-axis. The Gummel plot will provide us with many model parameters

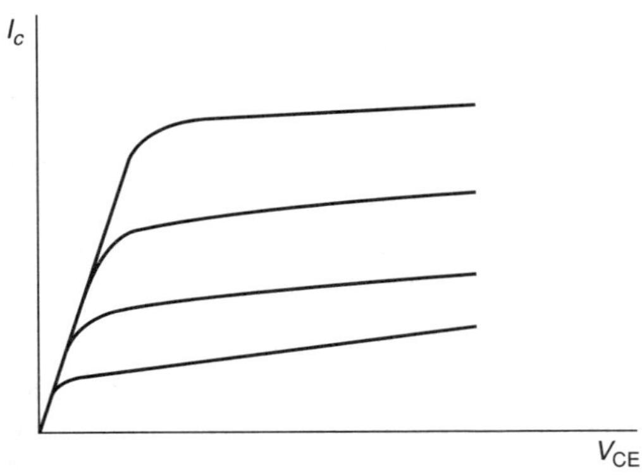

Figure 4.3(*a*) Bipolar family of curves.

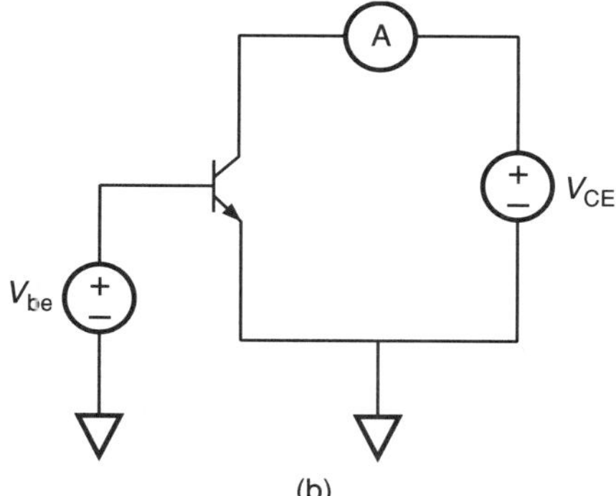

Figure 4.3(*b*) Measurement setup for family of curves.

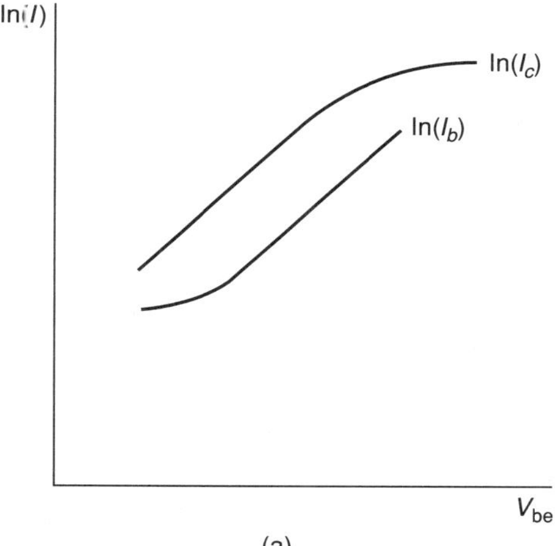

Figure 4.4(*a*) Bipolar Gummel curves.

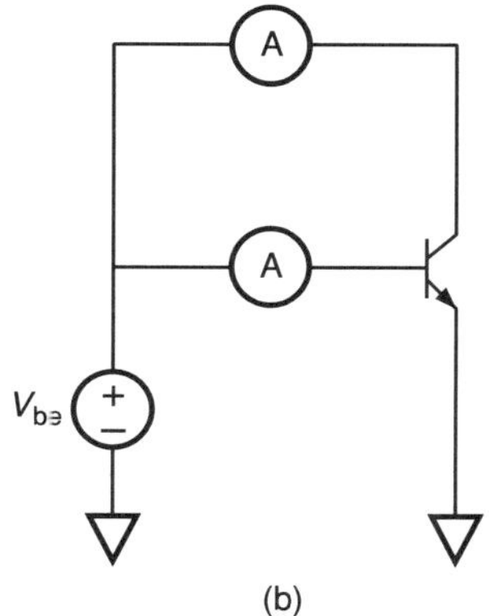

Figure 4.4(*b*) Measurement setup for Gummel curves.

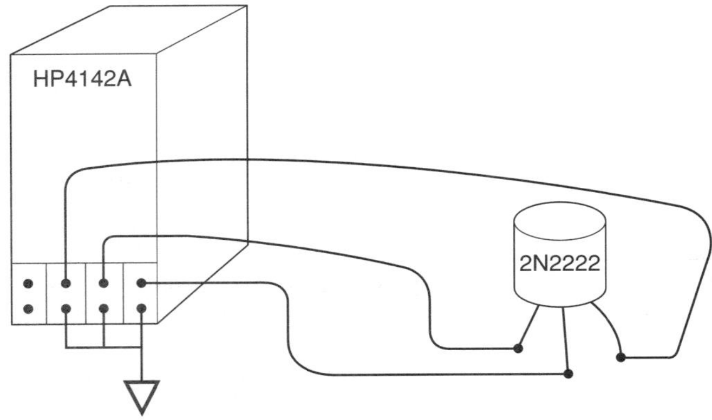

Figure 4.5 Test setup for Gummel curves.

of the bipolar device. *To model the bipolar transistor, we must be able to obtain the Gummel curves from the real device.*

To measure the Gummel characteristics from real transistors requires a parameteric analyzer. This measurement device was first mentioned in Chap. 3. A parametric analyzer is a fancy curve tracer with three or more programmable measurement units. The measurement units may be programmed to force current while measuring voltage, or force voltage while measuring current. Companies who supply parametric testers include Hewlett-Packard, Keithly Instruments, and Tektronix. For our tests, an HP 4142A parametric analyzer will be used. The test setup for the Gummel curves is shown in Fig. 4.5.

To obtain a plot of the Gummel characteristics, connect the bipolar transistor as shown in Fig. 4.6. Notice that, to obtain the correct Gummel curve, V_{bc} must be held at zero volts.

Simulating the Gummel curves is even easier than measuring the curves. To help you simulate the Gummel curves for our bipolar tran-

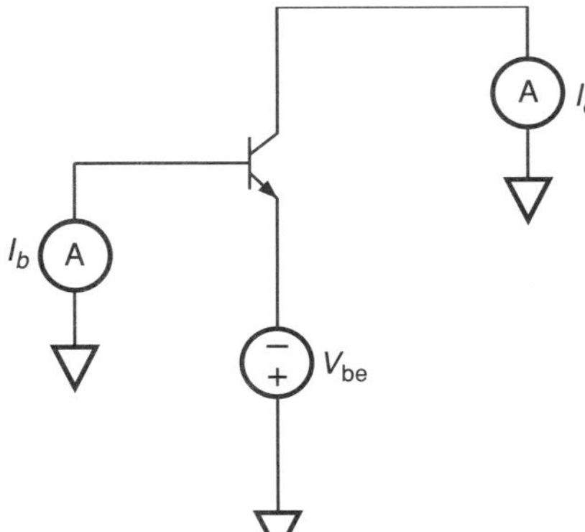

Figure 4.6 Measurement setup for Gummel curves.

sistor model, three simulation files have been created and placed in the CH4 subdirectory of the disk accompanying this text. After you have installed the disk on your computer according to the installation instructions, change into the CH4 subdirectory. In this subdirectory, you will find files named GUMMEL.BAT, GUMMEL.CIR, GUMMEL.MAC, and NPN.MOD. The GUMMEL.BAT file runs RSPICE on GUMMEL.CIR. The bipolar parameters in NPN.MOD are used during the simulation. The GUMMEL.CIR circuit file sweeps V_{be} while measuring the collector current (I(vic)) and the base current (I(vib)). When the simulation is complete, the batch file continues by running RGRAPH on the simulation output. The GUMMEL.MAC file is a macrocommand file which leads RGRAPH to plot the natural log of the base and collector currents.

To exercise this program, change into the `RCGV34\CH4` subdirectory and, at the DOS prompt, type:

```
GUMMEL
```

The resulting plot should match Fig. 4.7.

The GUMMEL.CIR circuit file reads the bipolar model parameters found in the NPN.MOD file. Changing the parameters of the NPN.MOD file will change the appearance of the resulting Gummel curves.

Also included in this same directory is the FAMILY.BAT file. This file will produce the family of curves described by the NPN.MOD

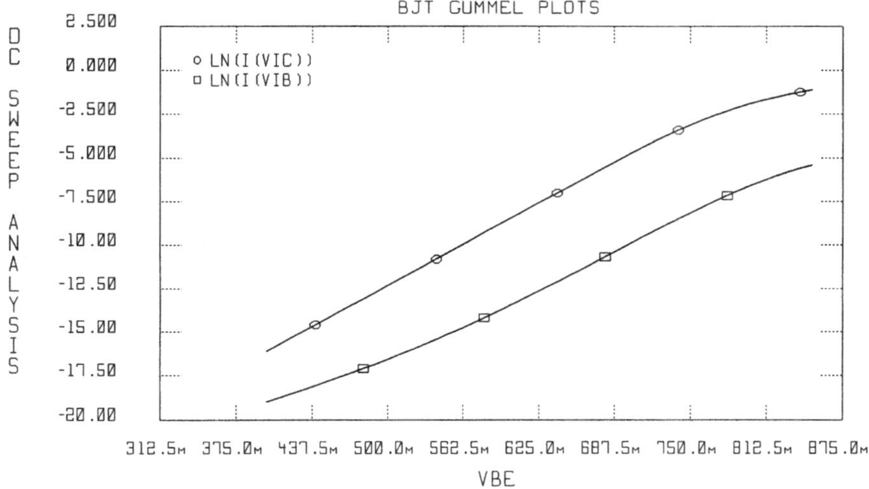

Figure 4.7 Gummel curve simulation output.

parameters. To exercise the FAMILY program, change into the `RCGV34\CH4` subdirectory and, at the DOS prompt, type:

```
FAMILY
```

The resulting plot should match Fig. 4.8.

Both of these curves will be important in modeling the bipolar transistor. As we develop model parameters from real devices, we will use these programs to simulate the I-V characteristics of the device to ensure the simulated model matches the real transistor.

We must return to the subject of extracting IS, BF, and NF for the transistor. To extract BF, we need merely measure the *maximum* vertical separation between the collector and base currents found on the Gummel curves. This measurement is best done as shown in Fig. 4.9. To determine the value of IS, a straight-line approximation should be drawn through the linear portion of the collector current curve. The intersection of the line and the y-axis is ln(IS). Use Eq. 4.3 to determine IS.

To determine BF, find the maximum vertical separation between the collector and base currents and its location. Measure the vertical separation between the two curves and calculate BF from Eq. 4.3.

To determine NF, measure the slope of the straight-line approximation used to find IS. NF of the bipolar model is equivalent to N for the diode model. Measure the slope of the straight-line approximation, and use Eq. 4.5 to determine NF.

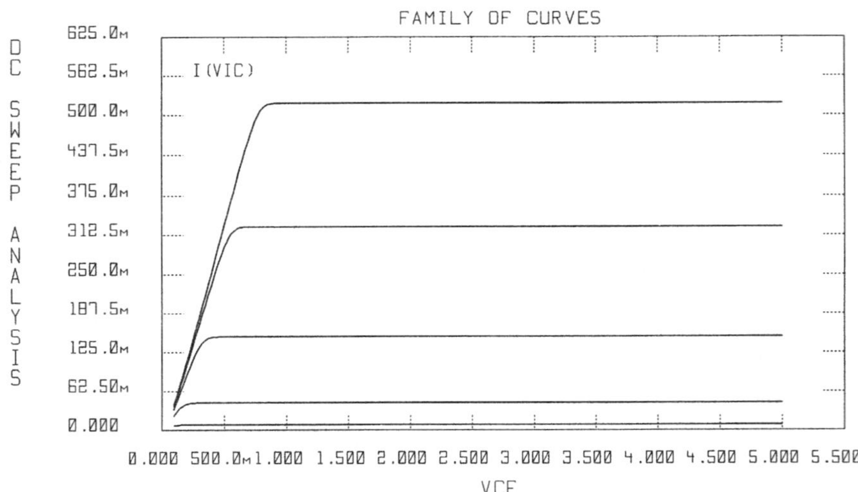

Figure 4.8 Family of curves simulation output.

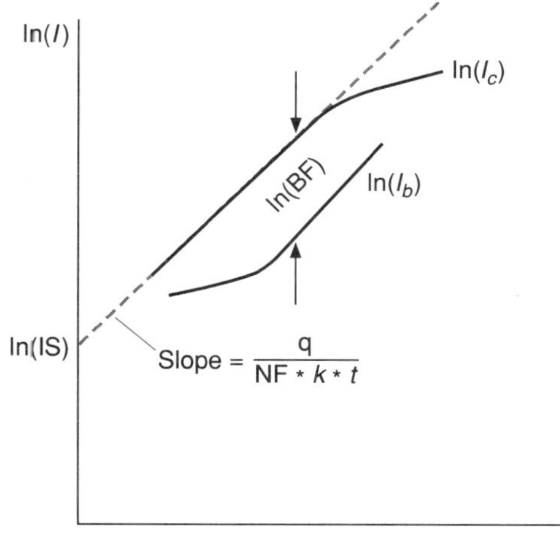

Figure 4.9 BF, IS, and NF extraction.

$$\text{IS} = \exp(y\text{-intercept of } \ln(I_c)) \tag{4.3}$$

$$\text{BF} = \exp(\ln(I_c) - \ln(I_b)) \text{ at point of maximum vertical separation} \tag{4.4}$$

$$\text{NF} = \frac{1}{\left(\text{slope} * \left(\frac{kT}{q}\right)\right)} \text{ or at room temp. NF} = \frac{1}{(\text{slope} * 0.0255)} \tag{4.5}$$

Since the forward beta or BF of the SPICE model represents the ratio (or the vertical distance between the two currents in Fig. 4.9) of collector current to base current, it should be evident from Fig. 4.9 that the forward beta of the model does not remain constant. At low base-emitter voltages, the base current is higher than classical physics would predict. This increase in base current is due to several nonideal effects, including recombination of carriers at the surface, recombination of carriers in the emitter-base depletion region, and the formation of emitter-base surface channels.[1] An increase in base current without a corresponding increase in collector current results in a reduction in the effective beta of the transistor.

At high V_{be} voltages, the slope of the collector current also rolls off. Again, because the reduction in collector current (I_c) is not accompanied with a corresponding reduction in base current, the beta at high currents is reduced. This reduction in beta at both high and low currents is evident if you plot beta or (h_{FE}) vs. I_c. A typical beta vs. I_c plot is shown in Fig. 4.10. The beta vs. I_c plot yields another method of determining the maximum current gain of the transistor BF. To deter-

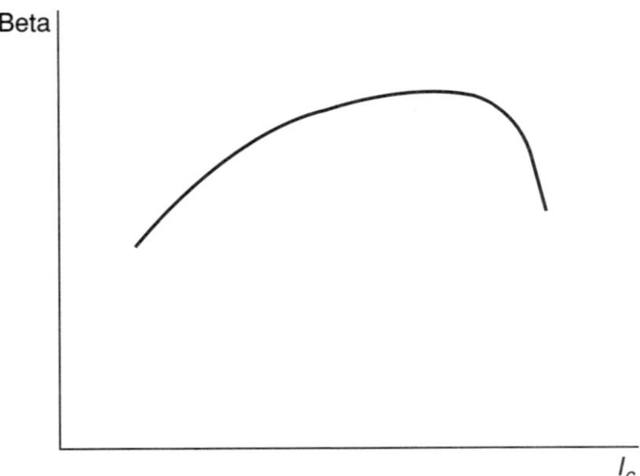

Figure 4.10 Beta vs. I_c curve.

mine BF from the beta vs. I_c plot, simply locate the maximum value of beta on the graph.

As we did with the Gummel curves, the beta vs. I_c curves may be obtained with a parametric analyzer. (In fact, the beta vs. I_c curves may be obtained once the data for the Gummel curves has been obtained.)

Simulating the beta vs. I_c curves is going to be an important means of verifying our model. To help you simulate the beta vs. I_c curves for any bipolar transistor, three simulation files have been created and placed in the CH4 subdirectory of the disk accompanying this text. After you have installed the disk on your computer according to the installation instructions, change into the `RCGV34/CH4` subdirectory. In this subdirectory, you will find files named BETA.BAT, BETA.CIR, BETA.MAC, and NPN.MOD. The BETA.BAT file runs RSPICE on BETA.CIR. The bipolar parameters in NPN.MOD are used during the simulation. The BETA.CIR circuit file sweeps V_{be} while measuring the collector current (I(vic)) and the base current (I(vib)). When the simulation is complete, the batch file continues by running RGRAPH on the resulting simulation. RGRAPH divides the collector and base current. The BETA.MAC is a macrocommand file which directs RGRAPH to plot beta vs. the collector current.

To exercise this program, change into the `RCGV34\CH4` subdirectory and, at the DOS prompt, type:

```
BETA
```

The resulting plot should match Fig. 4.11.

Table 4.2 illustrates the files which have been created for you to simulate the characteristics of your bipolar transistors.

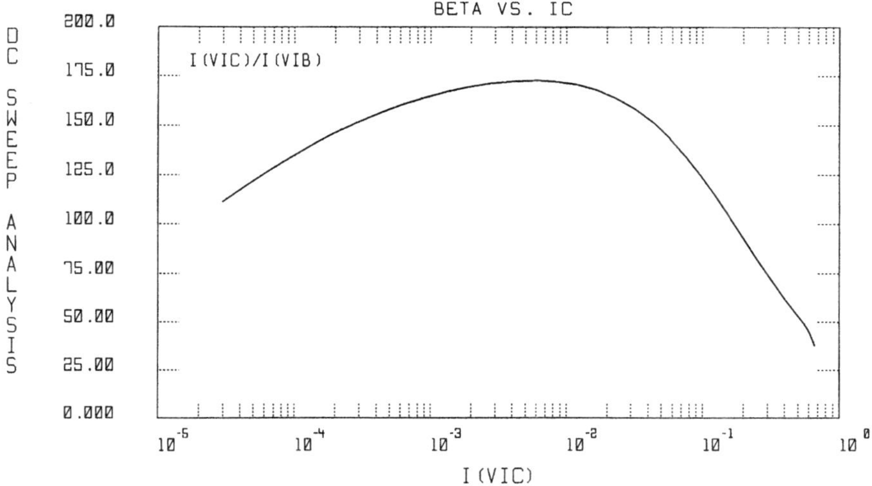

Figure 4.11 Simulated beta vs. I_c curve.

TABLE 4.2 Files for Simulating Bipolar Transistor Characteristics

GUMMEL.BAT	FAMILY.BAT BETA.BAT
GUMMEL.CIR	FAMILY.CIR BETA.CIR
GUMMEL.MAC	FAMILY.MAC BETA.MAC
NPN.MOD	

To model the reduction in beta (or BF) at both high and low base-emitter voltages requires three additional model parameters: ISE, NE, and IKF. [*Please note:* older versions of SPICE (prior to version 2f) used the parameter C2 in place of ISE. In these earlier versions of the program, C2 is equal to ISE/IS.] The additional base current which reduces the beta at low base-emitter voltages is modeled with ISE and NE. If a straight-line approximation is drawn through the nonideal portion of the base current curve, the intersection of that line on the y-axis is the natural log of ISE, and the slope of the line is inversely related to NE. Figure 4.12(a) illustrates the extraction of ISE and NE. Use Eq. 4.6 to determine ISE and Eq. 4.7 to determine NE.

$$\text{ISE} = \exp(y\text{-intercept}_{\text{IC Linear}}) \tag{4.6}$$

$$\text{NE} = \frac{1}{\left(\text{slope} * \left(\dfrac{kT}{q}\right)\right)} \text{ or at room temp} = \frac{1}{(\text{slope} \ *(0.0255))} \tag{4.7}$$

To model the high-current reduction in beta, we use the model parameter IKF. IKF is the value of collector current at which the slope of the collector current on the Gummel curve begins to fall. Figure 4.12(*a*) illustrates the extraction of ISE, NE, and IKF.

A word of caution about extracting IKF: IKF is only one of two mechanisms which may cause the collector current curve to deviate from the straight-line characteristics. The series resistance of the transistor RC and RE may cause the collector current to "roll off" from the ideal characteristics (in much the same way the series resistance of the diode made the diode curve roll off). The Gummel curves for a device with high series resistances is shown in Fig. 4.12(*b*). The Gummel curve characteristics cannot be used to determine whether the nonideal behavior is coming from the series resistance of the device or from IKF. For this reason, IKF will be determined *after* the series resistance model parameters have been extracted.

Figure 4.12(*a*) illustrates the importance of the Gummel plot in developing model parameters for the bipolar transistor. The six forward parameters, IS, NF, BF, ISE, NE, and IKF, are all measured directly from the Gummel plot. In addition, for transistors operated in reverse operation, the corresponding reverse parameters may be determined by reversing the polarity of the terminal voltages. This is shown in Fig. 4.13(*a*) with the required measurement setup in Fig. 4.13(*b*).

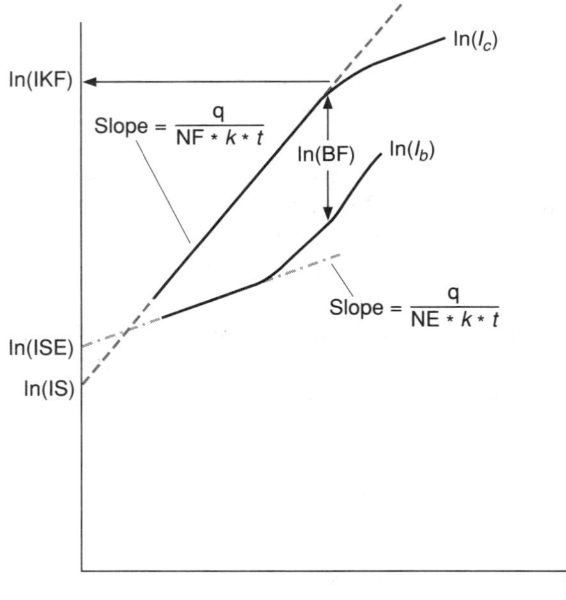

Figure 4.12(*a*) Bipolar Gummel curves.

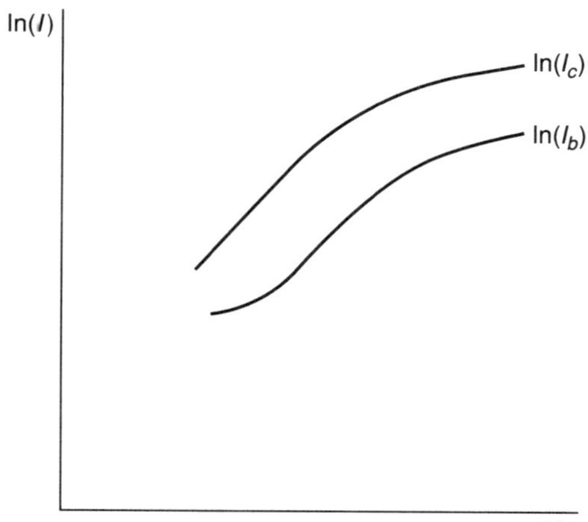

Figure 4.12(*b*) Gummel curves of transistor with large terminal resistances.

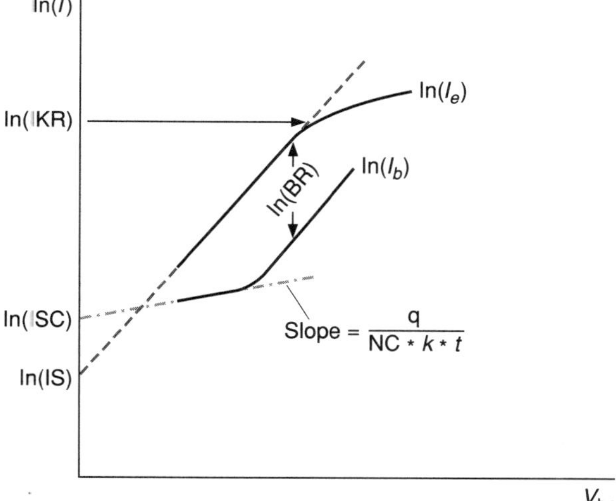

Figure 4.13(*a*) Reverse Gummel characteristics.

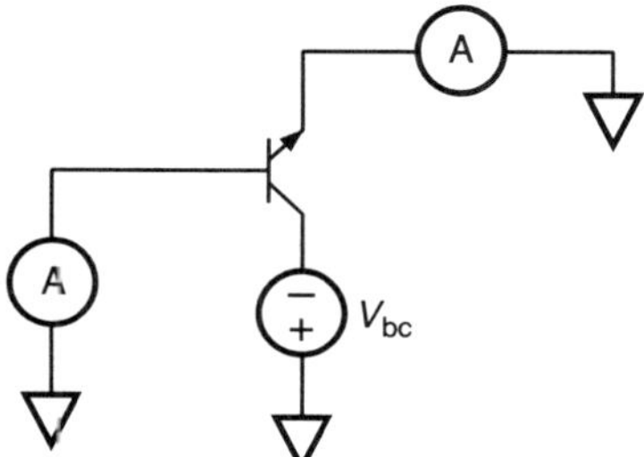

Figure 4.13(*b*) Measurement set-up for reverse Gummel test.

Without question, the Gummel parameters are the most important parameters of the bipolar model. Once these parameters have been determined, we can focus our attention on the next set of DC I-V parameters; the Early voltages VAF and VAR; and the emitter, collector, and base series resistances.

The Early voltage VAF (and VAR for reverse operation) models the base-width modulation effect of the transistor.[4] Base-width modulation causes the output conductance to be less than zero in the linear region of operation. The Early voltage, which is always entered as a positive number, is determined from the slope of the output characteristics of the family of curves, as shown in Fig. 4.14. (It should be noted that VAF was known as VA and VAR was known as VB in older versions of SPICE.) For reverse operation, the reverse Early voltage is determined as shown in Fig. 4.15.

Several nonideal behaviors of the bipolar transistor can be attributed to the series resistance of the collector, emitter, and base terminals. The model parameters which represent these resistances are RC, RE, and RB, respectively.

The emitter resistance is usually low since the emitter is normally doped with a high concentration of impurities. But because the product of the collector (or emitter) current times the emitter resistance generates a voltage which opposes the normal V_{be} voltage, the emitter resistance is an important parameter to model for bipolar devices. The emitter resistance can be determined by forcing a current into the base while measuring the open-circuit collector-emitter voltage. The test setup is shown in Fig. 4.16(*a*), and the resulting curve is shown in Fig. 4.16(*b*). The emitter resistance RE may be determined from the slope of the base-current curve. Use Eq. 4.8 to determine the emitter resistance.

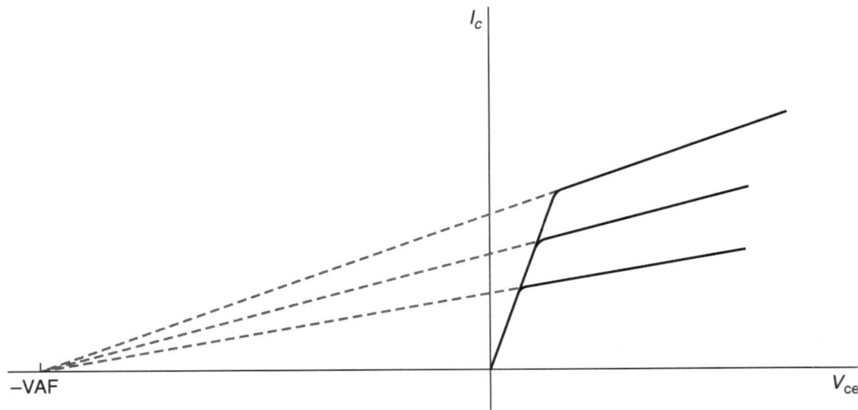

Figure 4.14 Forward Early voltage.

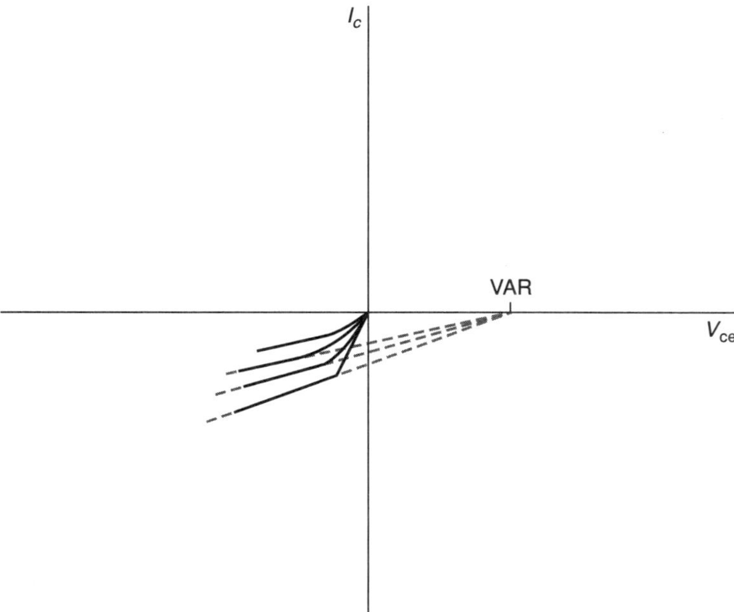

Figure 4.15 Reverse Early voltage.

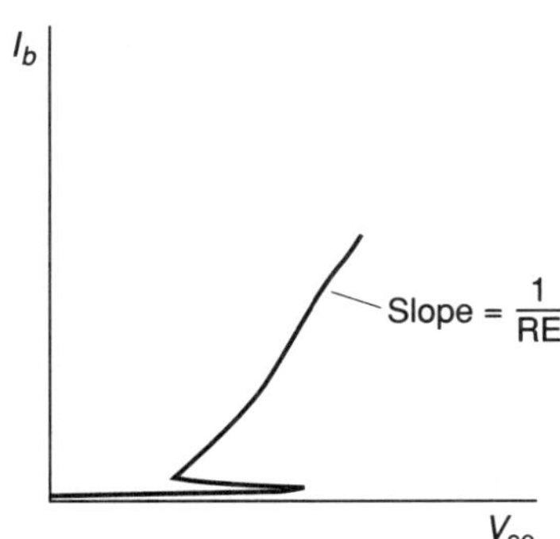

Figure 4.16(*a*) Emitter resistance extraction.

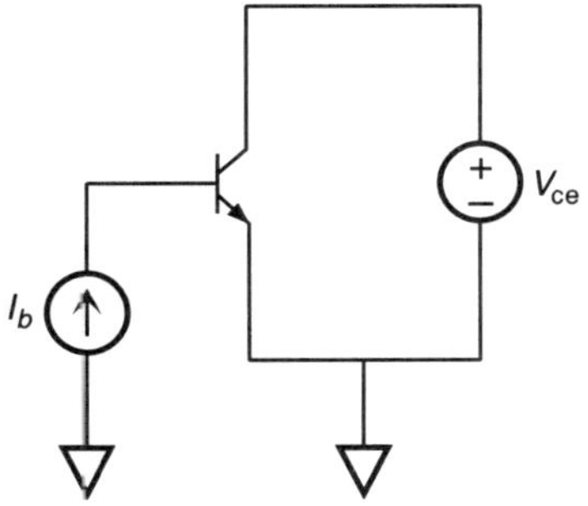

Figure 4.16(*b*) Measurement setup for emitter resistance test.

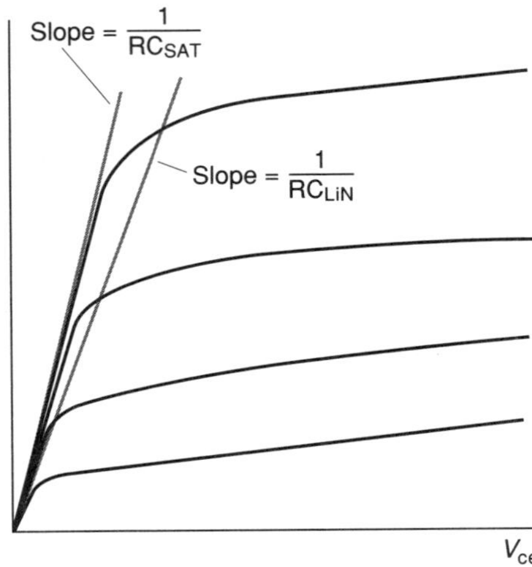

Figure 4.17 Collector resistance extraction.

$$\text{RE} = \frac{1}{\text{slope}(I_b)} \tag{4.8}$$

The collector resistance is slightly more difficult to determine, for while the collector resistance of the device will vary with the operating point of the transistor, the SPICE bipolar model forces us to use a fixed value of collector resistance. For this reason, the collector resistance value used in SPICE should be chosen depending on how the device is to be operated. Figure 4.17 shows the two limiting cases of collector resistance. The upper limit on collector resistance is known as RC_{LIN} and is determined by drawing a line through the knee of each curve of the family of curves. RC_{LIN} will be equal to the inverse of the slope of the straight line. If the transistor is mainly used in the linear region, the value of RC_{LIN} should be used for the model parameter RC. If a straight line is drawn through the linear portion of the I_c curve on the family of curves (with I_c at its highest value), a lower value of RC may be determined. The lower value of RC is known as RC_{SAT}. If the transistor is used for switching circuits or other applications where the device is used in saturation, the lower value of collector resistance RC_{SAT} should be used for the RC model parameter.

To determine RC, decide which region of operation will be used most often during simulation and set RC to either RC_{LIN} or RC_{SAT}. (In some cases, you may find that a value between these two limits produces the best results during simulation.)

Of the three terminal resistances, the base terminal resistance (RB) is the most difficult to determine. The reason for this is that the base

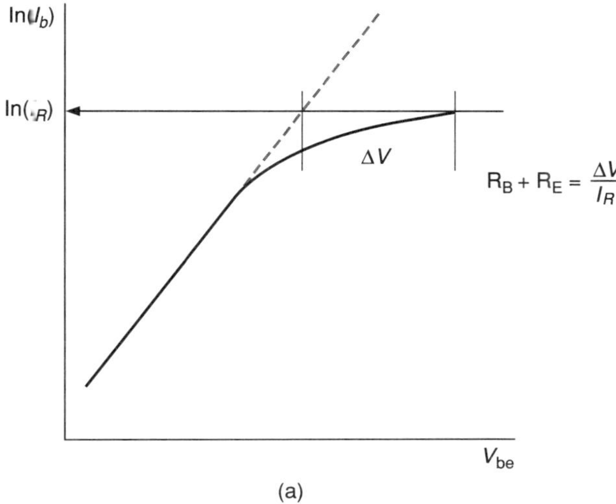

Figure 4.18(*a*) Base resistance extraction.

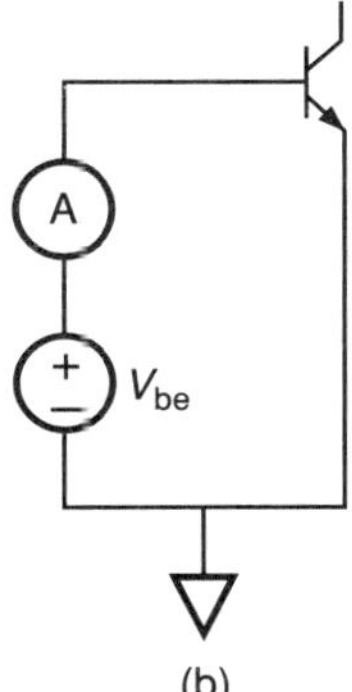

Figure 4.18(*b*) Measurement setup for base resistance extraction.

resistance changes with collector current of the transistor. In keeping with our philosophy of a *practical* transistor model, we will develop a test to determine RB *once the value of RE has been determined.*

To determine the base resistance we will perform a test similar to the test for series resistance of the diode. Connect the transistor to as shown in Fig. 4.18(*b*); sweep the base-emitter voltage while measuring the base current. Plotting the forward voltage (V_{be}) vs. the current on a logarithmic *y*-axis results in a plot similar to that shown in Fig. 4.18(*a*). From this plot, the series resistance (RB + RE) may be determined, and by subtracting RE from this value, RB can be found. RB may be found from Eq. 4.9.

$$RB = \frac{\Delta\_V}{I_r} - RE \tag{4.9}$$

The model parameters IRB and RBM may be used to model a variable base resistance; however, this kind of precision is beyond the scope of this book. The interested reader is referred to Gummel[3] and Getreu.[4]

With accurate values of the parameters IS, NF, BF, ISE, NE, IKF, VAF, RE, RC, and RB, users should have little if any problem modeling the DC I-V characteristics of the transistor. Once these parameters have been obtained, the user may turn their attention toward the capacitive and AC parameters of the transistor.

Capacitive and AC bipolar parameters

The capacitive and AC model parameters of the bipolar transistor include the junction capacitance parameters CJE, MJE, VJE, CJC, MJC, VJC, FC and, for integrated-circuit transistors with a substrate contact, CJS, MJS, and VJS, and possibly XCJC. Also important for an accurate AC model is the parameter TF and, for those dealing only with high-current devices, the parameters XTF, ITF, VTF, PTF, and TR may be important.

Of these parameters, the first 10 describe the junction capacitance of the base-emitter, base-collector, and, for integrated circuit transistors, the collector-substrate junction. (See Fig. 4.1(*a*) and (*b*).) Table 4.3 illustrates how these capacitances relate to the diode junction capacitance parameters we learned about in the last chapter. Many readers may find it surprising to know that SPICE uses the same junction capacitance equations for both the diode junction capacitance and the base-emitter, base-collector, and collector-substrate junctions of the bipolar transistor, but this revelation will simplify extracting these model parameter values. To extract the junction capacitance values of the bipolar, the base-emitter, base-collector, and the collector-substrate capacitance values must be measured against the reverse-bias junction voltage. The capacitance measurement is identi-

TABLE 4.3 Junction Capacitance Parameters of the Bipolar Transistor

Diode	Base-emitter	Base-collector	Collector-substrate
CJO	CJE	CJC	CJS
VJ	VJE	VJC	VJS
M	MJE	MJC	MJS
FC	FC	FC	FC

cal to the procedure used for the junction capacitance of a diode, only applied to the respective terminals of the bipolar transistor.

Once each of the respective junctions has been measured and the capacitance-voltage relation is known, the junction capacitance calculator CJ_CALC included in the UTIL subdirectory may be used to calculate the CJE, MJE, VJE, CJC, MJC, VJC, and CJS, MJS, and VJS parameters. FC will be left at the default value of 0.5 for all our simulations.

The parameter XCJC is probably going to be important only for integrated circuit transistors which operate at high frequencies.[1,4] XCJC*CJC represents the fraction of the base-collector junction capacitance on the internal base node to the collector. The remainder of the capacitance ((1 − XCJC) * CJC) is placed across the external base node to the collector. Since this parameter can be calculated only from the physical layout of the transistor, and since the parameter is important only for very high-frequency transistors, we will leave XCJC at its default value of 1 for our model. For readers interested in an exact value of XCJC, it is calculated from the layout geometry of the transistor. To determine XCJC, the physical area of the emitter and base diffusions must be known, then XCJC may be determined from Eq. 4.10.

$$\text{XCJC} = 1 - \left(\frac{\text{emitter area}}{\text{base area (including emitter)}}\right) \tag{4.10}$$

The forward transit time model parameter TF is used to model the excess charge stored in the base region when the base-emitter junction is forward-biased. For typical transistor operation, TF is constant, but at extremely high values of collector current, the value of TF rises above the low-current value. This effect is shown in Fig. 4.19.

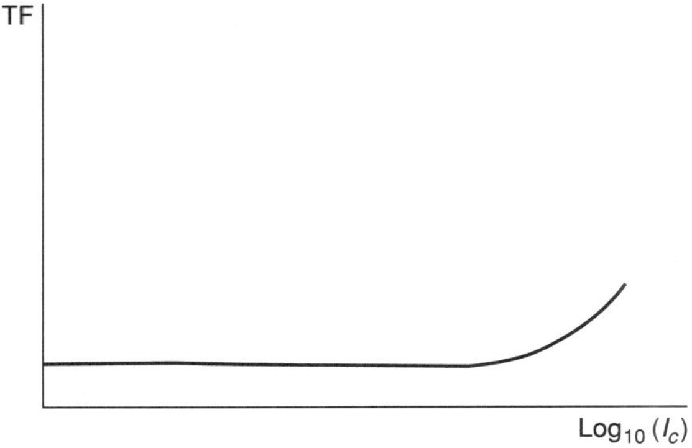

Figure 4.19 Transit time of bipolar transistor.

The high-current increase in TF is modeled with the parameters XTF, VTF, ITF, and PTF.

TF is best determined from f_T, the transistor's unity gain frequency. Many transistor data sheets contain a plot of the current gain (beta or h_{FE} vs. frequency), or the data sheet may simply give the value of the unity gain frequency. The frequency at which the gain of the transistor falls to unity represents the f_T frequency. If this information is absent from your data sheet, the unity gain frequency will have to be measured as shown in Fig. 4.20. Once the f_T frequency has been determined, TF may be determined from Eq. 4.11.

$$\mathrm{TF} = \frac{1}{(2 * \pi * f_T)} - \mathrm{CJC} * \mathrm{RC}_{\mathrm{LIN}} \tag{4.11}$$

For many small-signal transistors, the lone value of TF will provide an accurate model. Power transistors may require the additional parameters which will increase TF at high collector currents. Those additional parameters are XTF, VTF, PTF, and ITF.

Extracting the model parameters XTF, VTF, PTF, and ITF is difficult and unnecessary for many transistor models. For this reason, readers interested in extracting these parameter for very high current devices are directed to Gummel[3] and Getreu.[4]

The reverse parameter which corresponds to TF is TR. TR models the excess charge in the base-collector region when the base-collector junction is forward-biased. This condition implies reverse operation of the transistor and, for this reason, TR is probably important only for a device operated in reverse operation.

The extraction for TR is identical to the extraction for TF, except that the emitter and collector terminals must be swapped. Under these conditions, the current gain of the device is much lower than

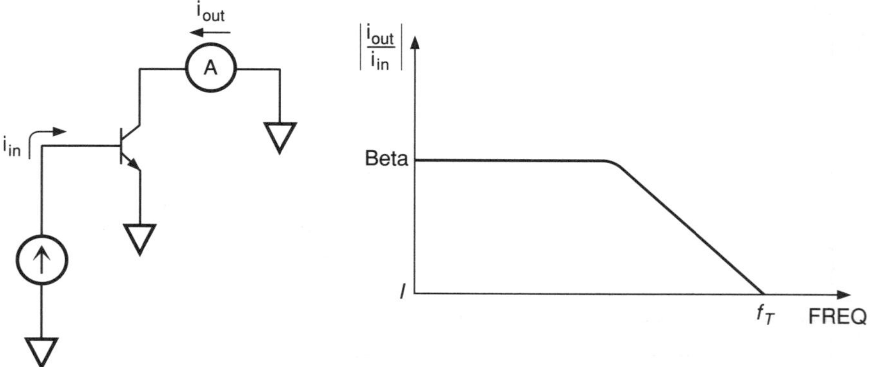

Figure 4.20 Measurement setup to determine f_T.

the normal forward current gain. The low reverse gain of the transistor results in a low value of f_T and a higher value of TR.

Bipolar Model Extraction Demonstration

In this section, we will determine the model parameters of a Motorola 2N2222a small-signal bipolar transistor.

Beginning the extraction requires an inspection of the data sheet. The first item which will be of use to us is the unity gain frequency f_T. On the data sheet, the minimum value of f_T is listed as 300 MHz. This value will be used to calculate TF, the forward transit time. Also on the data sheet is a plot of h_{FE} (BF in SPICE) vs. I_c.

What are notably absent from the data sheets are the all-important Gummel curves. Without the Gummel curves, our model is doomed. To produce an accurate model, we will have to generate our own Gummel plot. Also missing from the data sheet are plots of the junction capacitance vs. voltage for the base-emitter and base-collector junctions. Once again, we will measure the capacitance–voltage relationship for our transistors.

The data sheets for the 2N2222a transistor have loads of information for the designer, but very little for the device modeler. Again, the task of obtaining the necessary I-V and C-V characteristics of the device rests on the modeling engineer. We will have to make measurements on actual devices to do a good job of modeling the transistor. But isn't that what you have come to expect by now?

To begin the modeling process, we placed the transistor on our HP 4142 parametric analyzer, as shown in Fig. 4.21.

The first measurement made on the tester was the all-important Gummel test. The measurement setup is shown in Fig. 4.22. To produce the Gummel plots, the base and collector terminals are held at

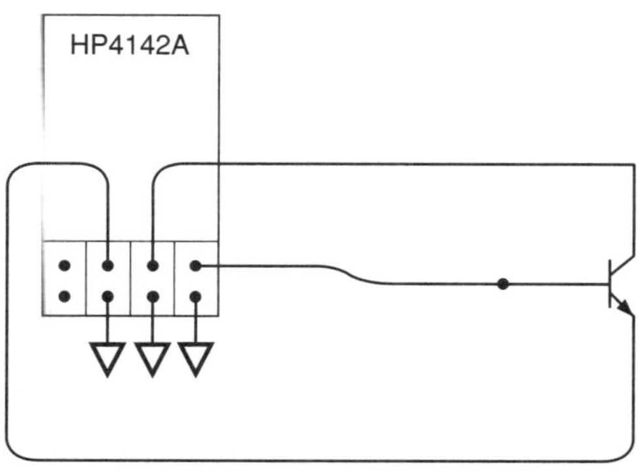

Figure 4.21 Test setup for Gummel test.

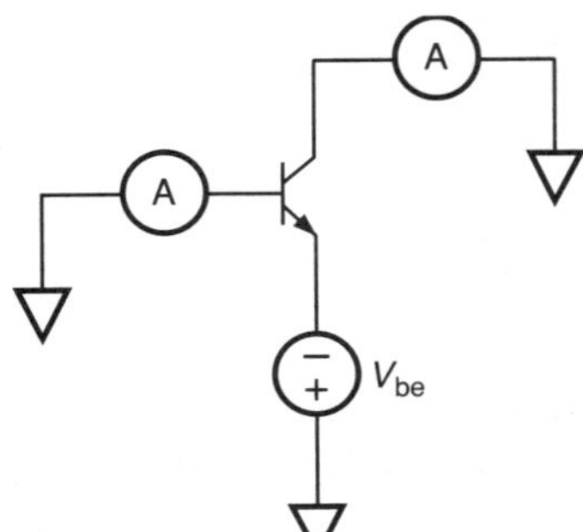

Figure 4.22 Measurement setup for Gummel test.

ground and the emitter is pulled to a negative voltage. This effectively produces a positive base-emitter voltage with the base-collector voltage held at zero volts. Both the base and collector currents are measured as the base-emitter voltage changes. The measured data was saved in an ASCII file and reproduced in RGRAPH. The collector current of the Gummel plot is shown in Fig. 4.23, the base current is shown in Fig. 4.24, and the complete Gummel plot is shown in Fig. 4.25.

Using the data measured from the Gummel test, the beta (h_{FE}) vs. I_c characteristics were constructed. The resulting beta vs. I_c plot is shown in Fig. 4.26.

The next measurement was the family of curves plot. The test setup for the family of curves is shown in Fig. 4.27. The family of curves may be created either by applying a base-current or a base-emitter voltage. For this test, the base-emitter voltage is swept from 0.6 to 0.9 V in 50-mV increments. The RGRAPH plot of the family of curves is shown in Fig. 4.28.

With the Gummel data in hand, we are ready to begin our extraction procedures. The first of the DC parameters we need are BF (h_{FE}), IS, and NF. All three of these values are obtained from the Gummel plot. IS is found from the intersection of I_c on the y-axis. BF is found from the maximum vertical distance between I_c and I_b (or the intersection of I_b on the y-axis). Finally, NF is determined from the slope of the collector current. All three of these extractions are shown in Fig. 4.29.

To extract IS, the y-intercept of the collector current must be found. This could be done graphically, as shown in Fig. 4.29, but much more accurate results are obtained by using a linear regression program to determine the intercept. In Chap. 3, the LINREG program was introduced. LINREG is a linear regression program that calculates the slope and intercept of a set of x and y data points. For those of you not familiar with the LINREG program, see App. C. The Gummel plot data was entered into the LINREG program. As shown in Fig. 4.29,

Vbe	Ic	ln(Ic)
0.40	8.58E-08	-16.2718
0.41	1.25E-07	-15.8918
0.42	1.83E-07	-15.5138
0.43	2.76E-07	-15.1036
0.44	4.06E-07	-14.7167
0.45	5.97E-07	-14.3307
0.46	8.73E-07	-13.9509
0.47	1.25E-06	-13.5956
0.48	1.85E-06	-13.1987
0.49	2.78E-06	-12.7945
0.50	4.13E-06	-12.3970
0.51	6.06E-06	-12.0145
0.52	8.87E-06	-11.6333
0.53	1.34E-05	-11.2217
0.54	1.97E-05	-10.8374
0.55	2.88E-05	-10.4537
0.56	4.14E-05	-10.0920
0.57	6.19E-05	-9.69015
0.58	9.06E-05	-9.30961
0.59	0.000136	-8.90286
0.60	0.000199	-8.52422
0.61	0.000287	-8.15673
0.62	0.000425	-7.76272
0.63	0.000623	-7.38064
0.64	0.000917	-6.99419
0.65	0.001342	-6.61359
0.66	0.001957	-6.23634
0.67	0.002818	-5.87173
0.68	0.004231	-5.46532
0.69	0.006187	-5.08530
0.70	0.008842	-4.72824
0.71	0.01258	-4.37565
0.72	0.01834	-3.99867
0.73	0.02564	-3.66360
0.74	0.03538	-3.34161
0.75	0.04782	-3.04031
0.76	0.06323	-2.76098
0.77	0.08366	-2.48099
0.78	0.103	-2.27303
0.79	0.1374	-1.98486
0.80	0.1647	-1.80363
0.81	0.1964	-1.6276
0.82	0.223	-1.50058
0.83	0.2668	-1.32126
0.84	0.2924	-1.22963
0.85	0.3342	-1.09602

Figure 4.23 Collector current for Gummel plot.

Vbe	Ib	Ln(Ib)	Vbe	Ib	Ln(Ib)
0.40	5.74E-09	-18.9756	0.63	3.45E-06	-12.5763
0.41	7.06E-09	-18.7694	0.64	5.03E-06	-12.1999
0.42	8.67E-09	-18.5636	0.65	7.29E-06	-11.8297
0.43	1.04E-08	-18.3824	0.66	1.01E-05	-11.501
0.44	1.35E-08	-18.1235	0.67	1.58E-05	-11.0555
0.45	1.65E-08	-17.9199	0.68	2.23E-05	-10.7114
0.46	2.06E-08	-17.6965	0.69	3.25E-05	-10.3352
0.47	2.58E-08	-17.4717	0.70	4.70E-05	-9.96515
0.48	3.23E-08	-17.2473	0.71	6.85E-05	-9.58926
0.49	4.12E-08	-17.0048	0.72	0.000107	-9.14268
0.50	5.36E-08	-16.7421	0.73	0.000145	-8.84016
0.51	6.93E-08	-16.4844	0.74	0.000210	-8.47031
0.52	9.16E-08	-16.2055	0.75	0.000294	-8.13057
0.53	1.24E-07	-15.9062	0.76	0.000423	-7.76861
0.54	1.64E-07	-15.6258	0.77	0.000598	-7.42242
0.55	2.25E-07	-15.3054	0.78	0.000815	-7.11245
0.56	3.02E-07	-15.0122	0.79	0.001135	-6.78112
0.57	4.26E-07	-14.6693	0.80	0.001468	-6.52385
0.58	5.99E-07	-14.3287	0.81	0.001939	-6.24558
0.59	8.35E-07	-13.9955	0.82	0.002461	-6.00719
0.60	1.15E-06	-13.6775	0.83	0.003041	-5.79557
0.61	1.64E-06	-13.3214	0.84	0.003797	-5.57354
0.62	2.47E-06	-12.9129	0.85	0.004432	-5.4189

Figure 4.24 Base current for Gummel plot.

only the values of I_c that produced a linear characteristic should be used for the regression. Entering current values in the nonlinear region will distort the output of the LINREG program. The Gummel data used in the LINREG program to calculate IS and NF is shown in Table 4.4. Running LINREG produces the results seen in Fig. 4.30.

From Fig. 4.30, the intersection of the collector current on the y-axis is −31.7. To determine IS, we need to enter this value in Eq. 4.3.

$$\text{IS} = \exp(-31.7) = 17.09\text{ e }-15\text{ A} \quad \text{or} \quad 0.01709\text{ pA} \tag{4.12}$$

The slope of the Gummel curve is used to determine NF. From Fig. 4.30, the slope of the collector current is 38.59. Since the measurement was done at room temperature, we can use the simplified form of Eq. 4.5.

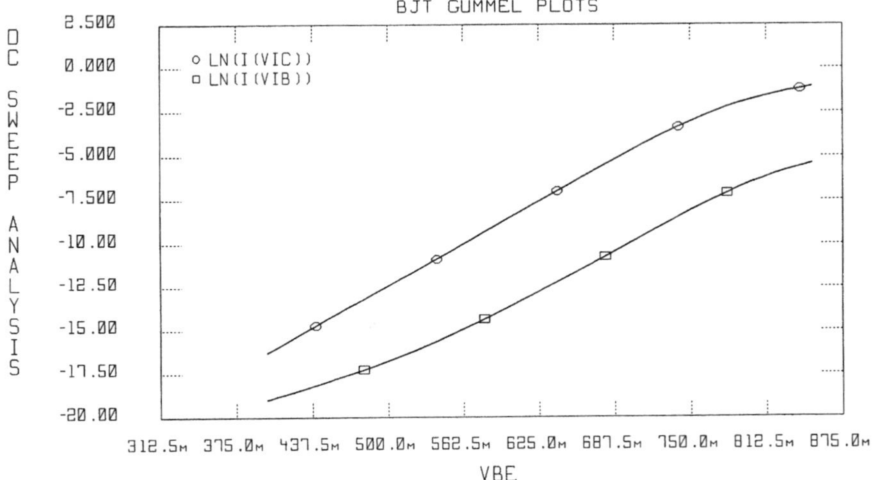

Figure 4.25 Gummel plot of 2N2222a transistor.

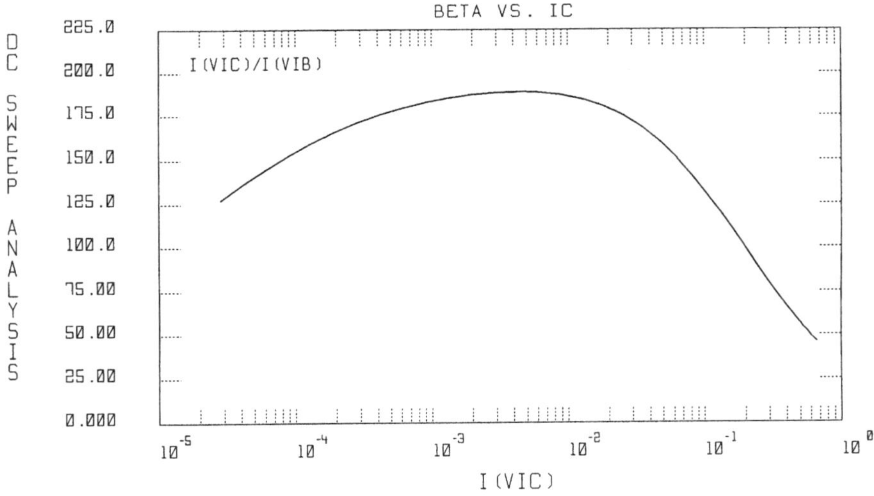

Figure 4.26 Beta vs. I_c for 2N2222a transistor.

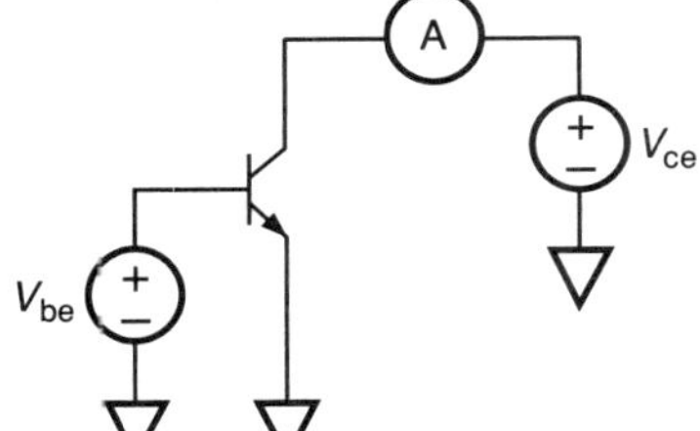

Figure 4.27 Measurement setup for family of curves.

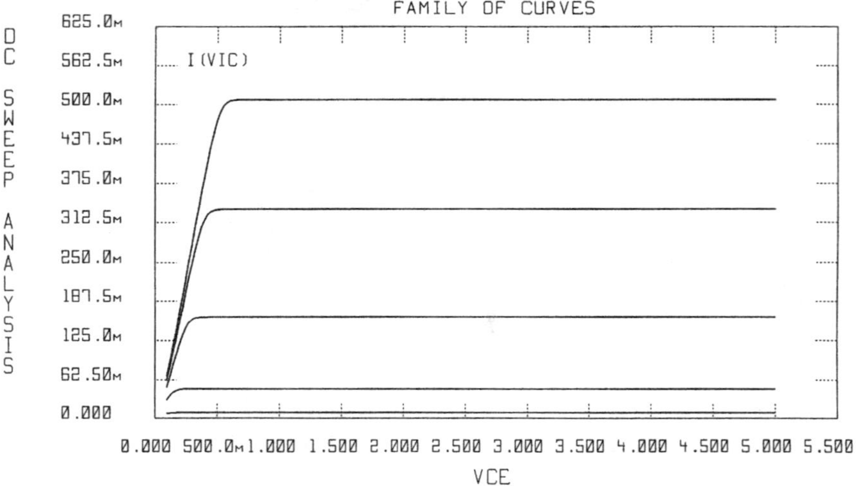

Figure 4.28 Family of curves from 2N2222a transistor.

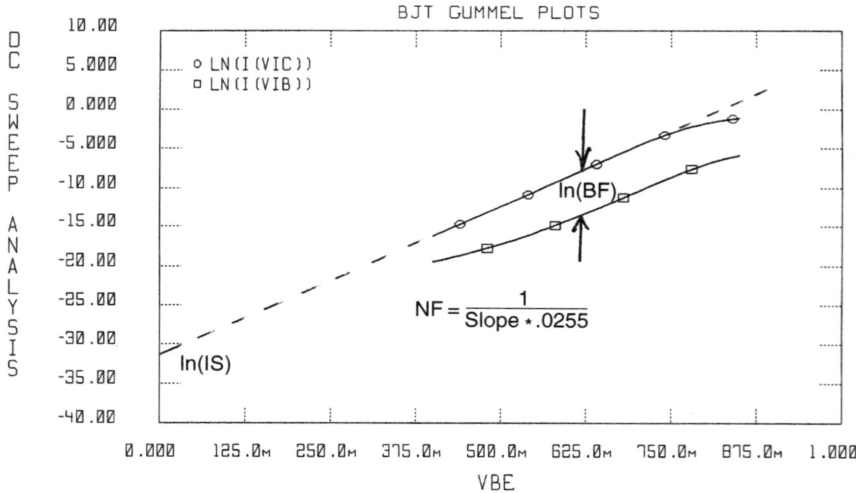

Figure 4.29 Extracting IS, BF, and NF.

$$\text{NF} = \frac{1}{(38.59 * 0.0255)} = 1.02 \tag{4.13}$$

To determine BF (h_{FE}) we can either search the Gummel curve for the maximum vertical separation between I_c and I_b or we can simply turn to the beta-vs.-I_c plot and read the maximum from the curve. The latter is probably the easiest. From Fig. 4.31, the maximum beta is 188.47.

TABLE 4.4 Gummel Data Used to Calculate IS and NF

V_{be}	$\ln(I_c)$
0.4	−16.2718
0.41	−15.8918
0.42	−15.5138
0.43	−15.1036
0.44	−14.7167
0.45	−14.3307
0.46	−13.9509
0.47	−13.5956
0.48	−13.1987
0.49	−12.7945
0.5	−12.397
0.51	−12.0145
0.52	−11.6333
0.53	−11.2217
0.54	−10.8374
0.55	−10.4537
0.56	−10.092
0.57	−9.69015
0.58	−9.30961
0.59	−8.90286
0.6	−8.52422
0.61	−8.15673
0.62	−7.76272
0.63	−7.38064
0.64	−6.99419
0.65	−6.61359
0.66	−6.23634
0.67	−5.87173
0.68	−5.46532
0.69	−5.0853
0.7	−4.72824

```
* * * * * * * * * * LINREG Results * * * * * * * *
*
* For the equation Y = a + b*X
*
* a    =   -0.31702E+02
* b    =    0.38593E+02
* chi2 = 0.009
*
```

Figure 4.30 LINREG output from V_{be} vs. $\ln(I_c)$

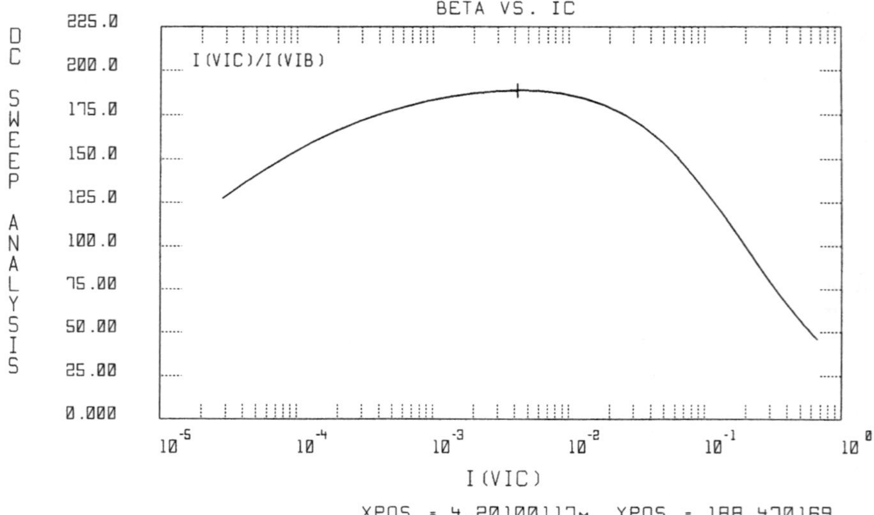

Figure 4.31 Beta vs. I_c.

With BF determined, we are ready to begin construction of our model. To this point, we have determined the values shown in Table 4.5. These values were entered into the NPN.MOD file in the CH4 subdirectory. By entering these values in this file, we can simulate the Gummel curves, the family of curves, and the beta-vs.-I_c curve. Figure 4.32 illustrates the simulated model and the measured data to this point.

Close inspection of Fig. 4.32 demonstrates that the modeled base and collector currents are slightly less than the actual device. The value of IS represents the offset of these curves on the Gummel plot. From Fig. 4.32, it appears that the value of IS is too low (the modeled characteristics need to be raised) for our model. The low value of IS is probably due to the selection of data points used in the LINREG program. If you look carefully at Fig. 4.32, you will see that the collector current curve actually has a slight curvature to it, with V_{be} between 0.4 and 0.5 V. Since the data used in the LINREG program included V_{be}s from 0.4 to 0.7 V, the resulting intercept and slope was the best

TABLE 4.5 Extracted Bipolar Model Parameters

Parameter name	Value
BF	188
IS	17.09 fA
NF	1.02

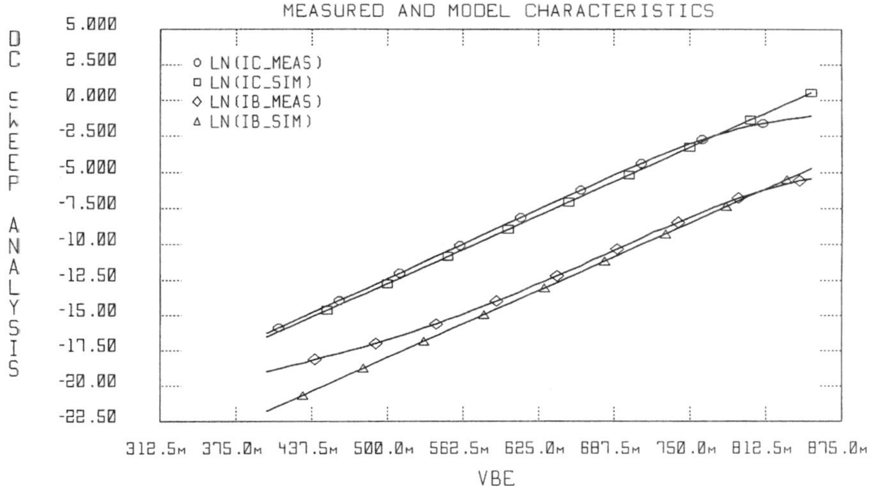

Figure 4.32 Gummel plot comparison after preliminary extraction of IS, NF, and BF.

fit of all data. This results in a value of IS which is too low for our model.

While we could rerun the LINREG program over a limited range of the data, this author would prefer to be *practical* about the solution. Rather than rerunning the LINREG program, an improved value of IS can be determined by simply noting that the simulated characteristics are offset by the measured characteristics by what appears to be a constant amount. By measuring the y-axis offset, as shown in Fig. 4.33, the offset can be added to the previously determined value of y-intercept and used to determine a new value of IS. The measured value of y-axis offset from Fig. 4.33 is shown to be approximately 0.43.

The calculation is of the new value of IS is illustrated in Eq. 4.14.

$$\text{IS} = \exp(-31.7 + 0.43) = 26.28 \text{ fA} \tag{4.14}$$

The new value of IS was added to our model, and the model was resimulated and compared with the measured values. The result of this simulation is shown in Fig. 4.34.

The next parameters needed for the model are the series resistance parameters. Remember, the series resistance parameters must be determined before the rest of the Gummel parameters, because the effect of these two parameters can produce similar characteristics on the Gummel curves.

To determine RC, we need to determine the maximum and minimum slope of the collector current in the linear region. The maximum

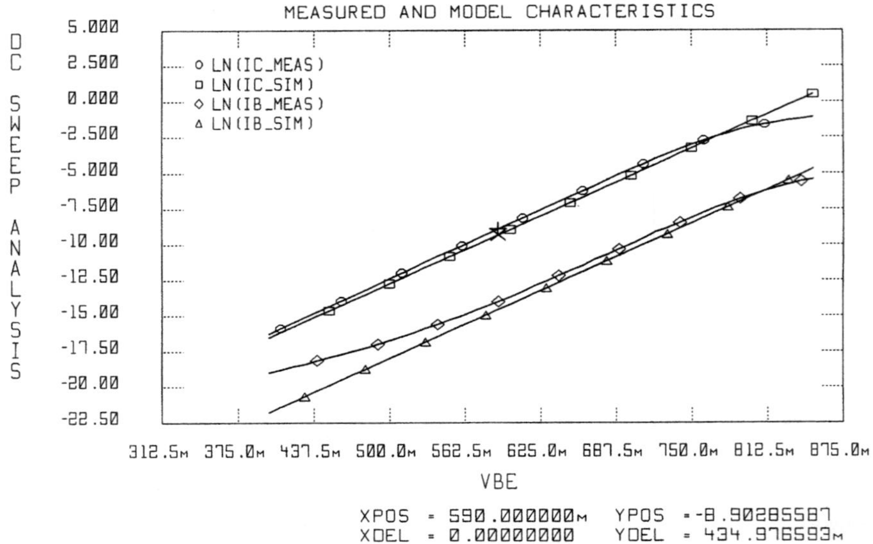

Figure 4.33 Estimating offset between simulated and measured curves.

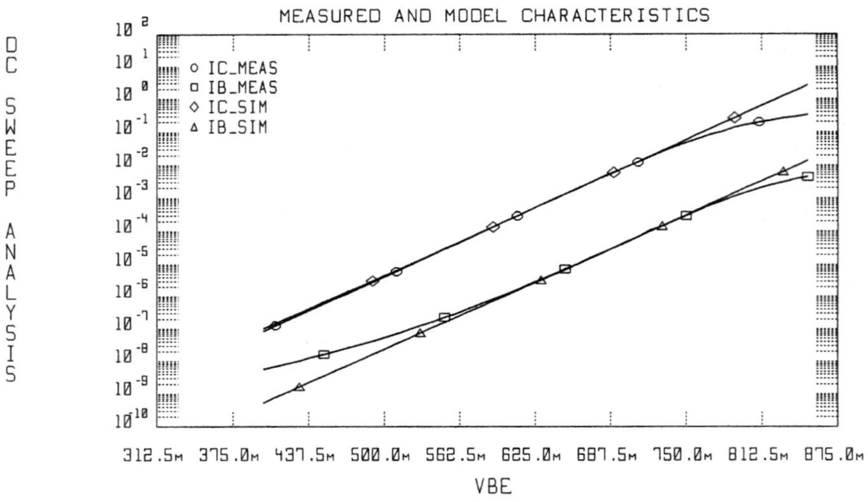

Figure 4.34 Gummel curves after optimizing value of IS.

slope will be RC_{SAT}; the minimum will be RC_{LIN}. The value we choose for the model will fall somewhere between these limits.

The measured family of curves is shown in Fig. 4.35(*a*) and (*b*). While we could use the LINREG program to determine the slope of the collector current in the saturated region, it is much easier to simply use the cursor and marker functions in RGRAPH to determine the maximum and minimum slope in the linear region. Figure 4.35(*a*) illustrates the position of the markers in preparation for determining the maximum slope, while Fig. 4.35(*b*) illustrates the markers for the minimum slope calculation.

The maximum slope in the saturated region is used to determine RC_{SAT} and the minimum slope is used to determine RC_{LIN}. Equations 4.15 and 4.16 illustrate the determination of the limiting cases of RC.

$$RC_{LIN} = \frac{1}{\text{minimum slope}} = 1.4\ \Omega \tag{4.15}$$

$$RC_{SAT} = \frac{1}{\text{maximum slope}} = 1.1\ \Omega \tag{4.16}$$

For our model, we will begin by setting RC equal to RC_{SAT}. If the transistor was used only in switching applications or is always biased in the linear region, RC should be set to RC_{LIN}.

Following the collector resistance test, we are set to determine the emitter resistance. The emitter resistance test consists of injecting a current into the base of the transistor while measuring the open-circuit collector-emitter voltage. The test setup is shown in Fig. 4.36(*a*) and the result is shown in Fig. 4.36(*b*). The slope of the resulting curve (in the linear region between the markers) is used to determine the emitter resistance. The resistance is calculated with Eq. 4.17.

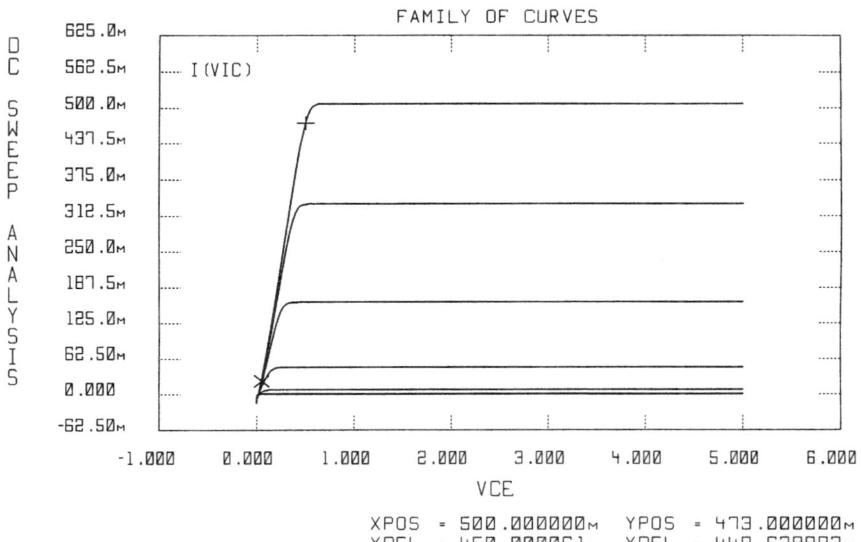

Figure 4.35(*a*) Setting markers to determine maximum slope for collector resistance extraction (RC_{SAT}).

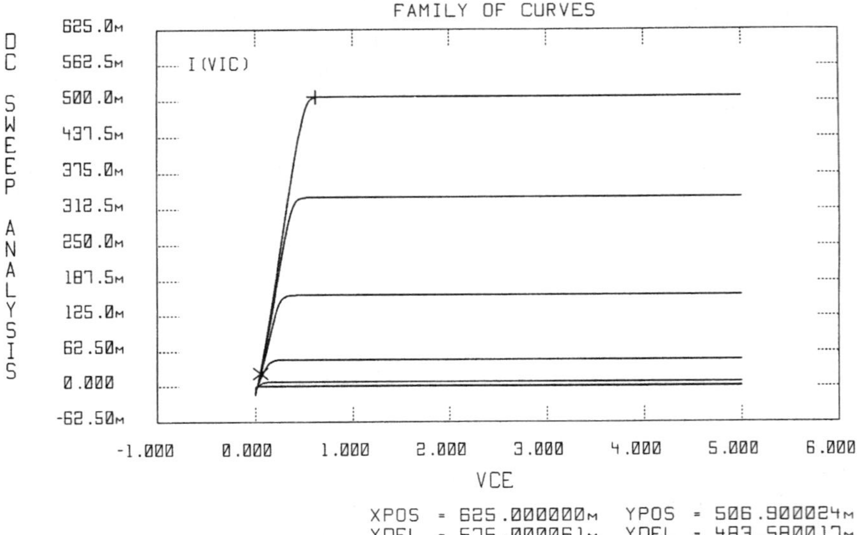

Figure 4.35(*b*) Setting markers to determine minimum slope for collector resistance extraction (RC_{LIN}).

$$\mathrm{RE} = \frac{1}{\mathrm{slope}(I_b)} = \frac{1}{(38.37)} = 0.026\ \Omega \tag{4.17}$$

With the emitter resistance determined, we can move on to the base resistance calculation. The base resistance calculation follows the theory of the diode series resistance tests. As shown in Fig. 4.37(*a*), the forward voltage of the base-emitter junction is swept while the base current is measured. The resulting I-V plot is similar in appearance to a diode curve, and when plotted on a logarithmic axis, appears as illustrated in Fig. 4.37(*b*). As with the diode series resistance test, the voltage difference between the ideal (linear curve characteristics and the actual (nonlinear curve) at the highest value of current determines the resistance of the device. In this case, the resistance of the

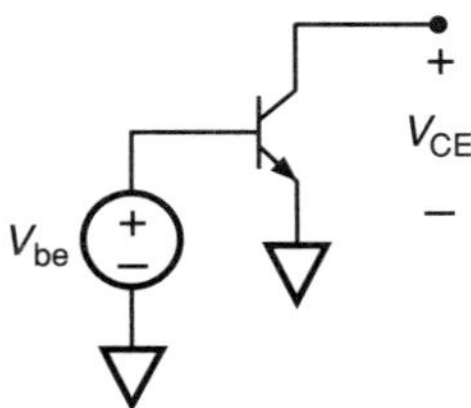

Figure 4.36(*a*) Measurement setup for emitter resistance extraction.

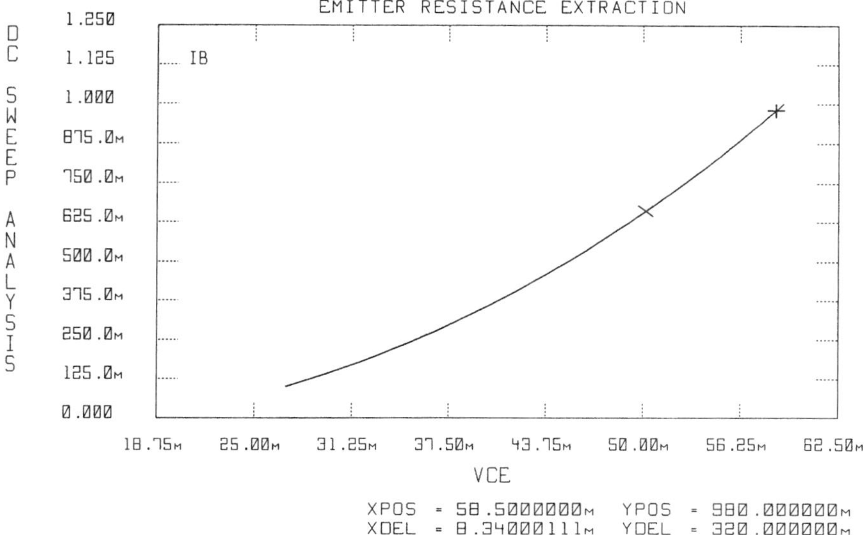

Figure 4.36(*b*) Plot of base current vs. V_{ce} voltage.

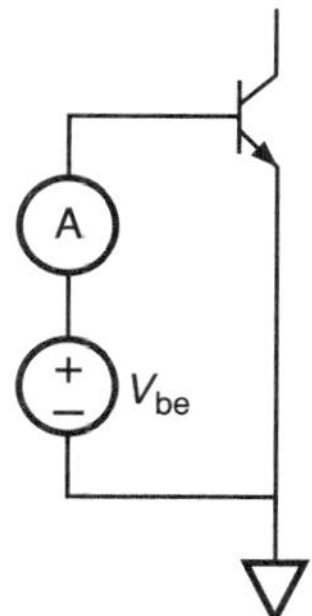

Figure 4.37(*a*) Measurement setup for base resistance extraction.

device includes both the base resistance and the emitter resistance. Since the emitter resistance has been previously determined, the base resistance may be found from Eq. 4.18.

$$RB = \frac{(0.886\text{v} - 0.75\text{V})}{30\text{ mA}} - RE = 4.533 - 0.026 = 4.5\Omega \quad (4.18)$$

Table 4.6 illustrates the three terminal resistance values for our model. Adding the terminal resistances to our model, and comparing the new simulation results with the measured data results in the graph of Fig. 4.38. This graph shows some improvement over the graph in Fig. 4.34, but both the measured high collector current and low base currents disagree with modeled currents.

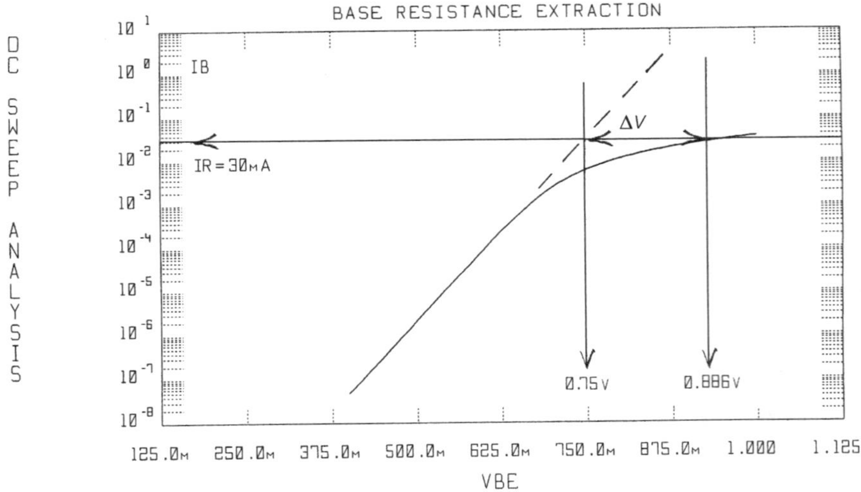

Figure 4.37(*b*) Base resistance extraction.

TABLE 4.6 Terminal Resistance Values for the 2N2222a Model

Resistance name	Value
RB	4.5
RE	0.026
RC	1.1

The high collector current roll-off is due to high-level injection effects and is modeled in SPICE with the parameter IKF. IKF represents the collector current where the roll-off from the ideal current begins. From Fig. 4.39, it appears that the measured current starts to deviate from the simulated current at $\ln(I_c) = -2.5$, or at $I_c = \exp(-2.5) = 82$ mA. Setting IKF to 82 mA results in Fig. 4.40(*a*). In this graph, the modeled characteristics improve, but the simulated collector current rolls off a little too soon. To try to improve on the model, several simulations were done with increased values of IKF. In these simulations, IKF was set to 100, 160, and 200 mA. Of these, the 200 mA proved to be the best fit for the model. The final simulation, with IKF set to 200 mA, is shown in Fig. 4.40(*b*).

This example of changing IKF, simulating, and comparing the simulated and measured characteristics to pick an optimum IKF value illustrates why optimization has become such a popular way to develop model parameter sets. Model optimizers read measured I-V data, make an initial guess for the model parameter values, then simulate

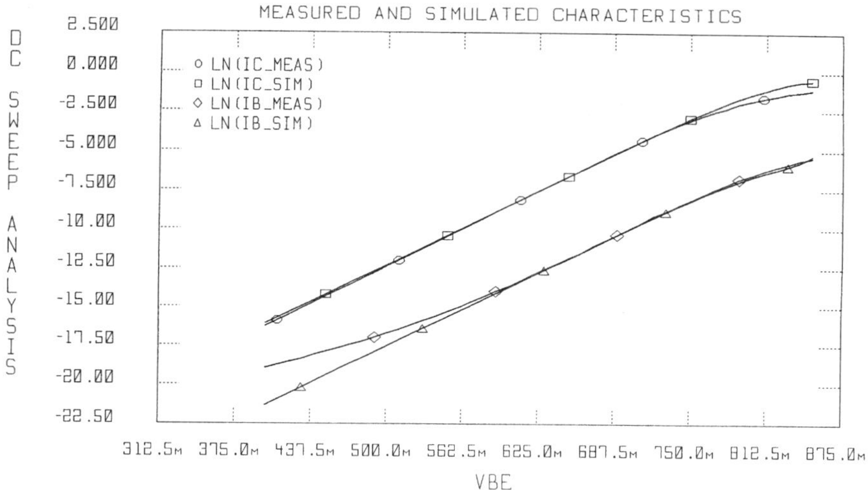

Figure 4.38 Gummel curves after RB, RC, and RE extraction.

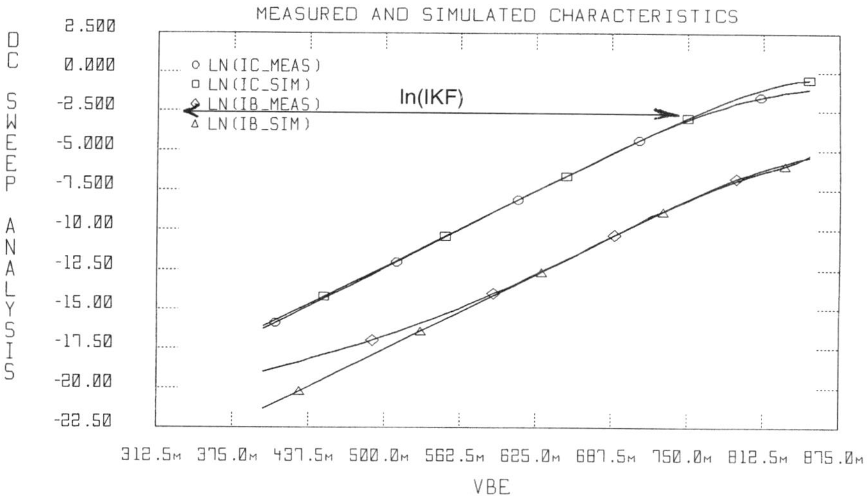

Figure 4.39 Extracting IKF.

the model over and over while changing the parameter values. The resulting model parameters produce a simulated characteristic very similar to the measured data. But software optimizers are very expensive, with most starting at about $35,000, out of the reach of many simulation users. (It is the author's hope that this text, which will probably sell for about 1/700th the cost of an optimizer, will help you produce similar results.) But by knowing what the parameters do to the simulated curves, and by knowing how to make an educated guess to the initial value of a parameter, SPICE users can manually optimize parameter values to produce an accurate model, too.

Figure 4.40(b) illustrates that the accuracy of our model is good for both high and mid-range values of V_{be}, but at low V_{be} voltages, the base current is unacceptably inaccurate. The model predicts a base

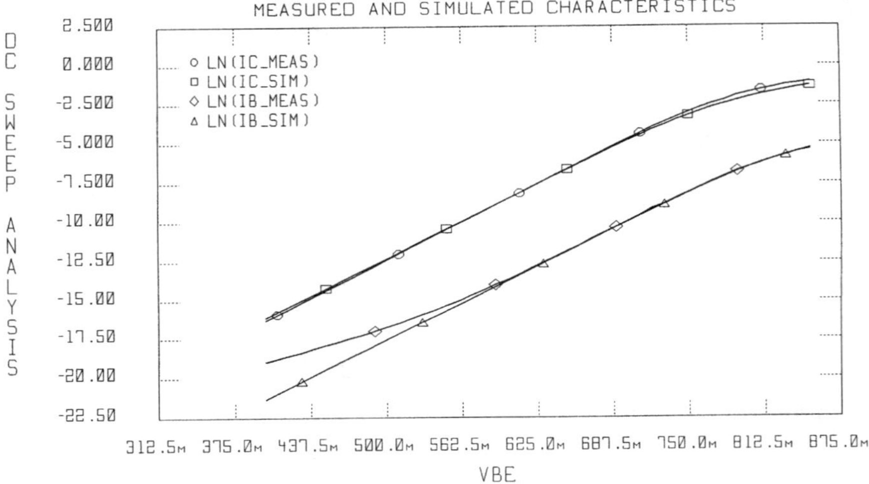

Figure 4.40(*a*) Comparison of simulated and measured characteristics with IKF set to 82 mA.

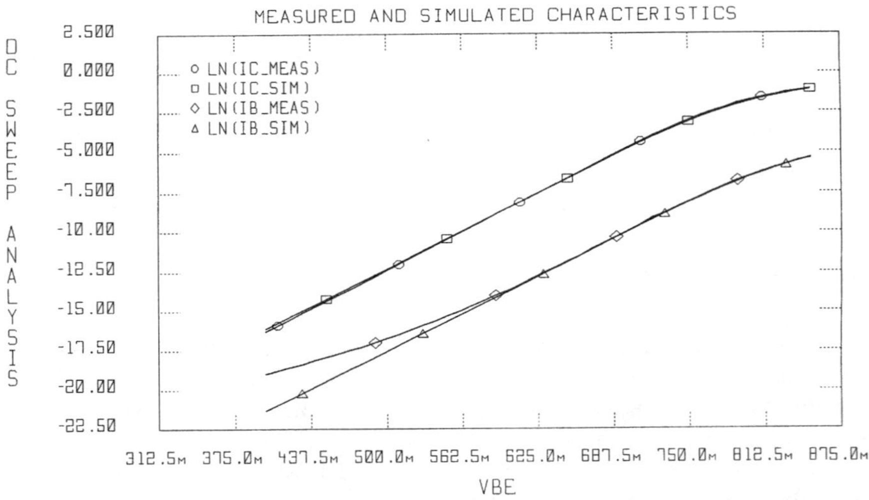

Figure 4.40(*b*) Comparison of simulated and measured characteristics with IKF set to 200 mA.

current which is much lower than the measured base current. The measured base current is higher than the simulated characteristic because of recombination of the carriers in the base region and at the surface of the base region. The SPICE model parameters which account for this nonideality are ISE and NE.

As illustrated in Fig. 4.12, ISE is the intersection on the y-axis of the Gummel curve, and NE is calculated from the slope of the nonlinear region. To determine both of these, data was taken from the Gummel curves in the low V_{be} region and entered into the LINREG program. The data entered into the LINREG program is shown graphically in Fig. 4.41 and numerically in Table 4.7. The output of the LINREG program is shown in Fig. 4.42. ISE and NE are calculated as shown in Eqs. 4.19 and 4.20.

$$\text{ISE} = \exp(y\text{-intercept}) = \exp(-27.21) = 1.5 \text{ pA} \tag{4.19}$$

$$\text{NE} = \frac{1}{(\text{slope}^*(kT/q))} \quad \text{or at room temperature} \quad \frac{1}{(\text{slope}^*(0.0255))} \tag{4.20}$$

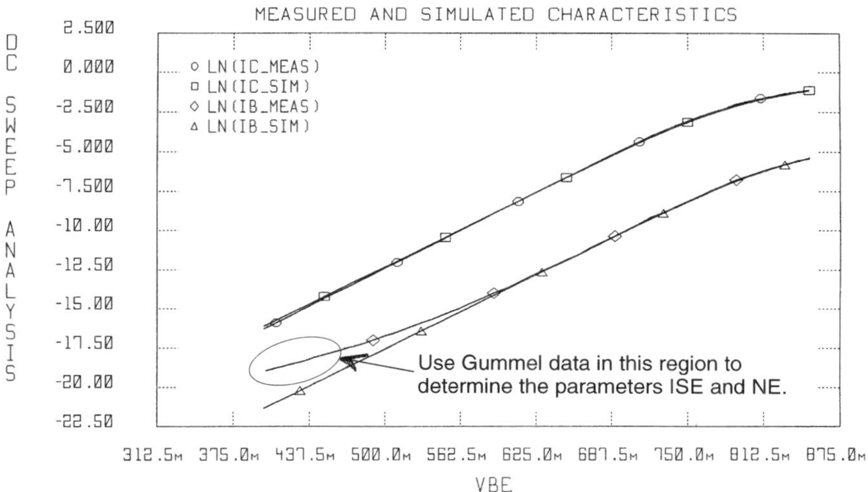

Figure 4.41 Extracting ISE and NE from Gummel curves.

TABLE 4.7 Gummel Data for Extraction of ISE and NE

V_{be}	$\ln(I_b)$
0.40	−18.97
0.41	−18.76
0.42	−18.56

$$\text{NE} = \frac{1}{(20.6^*0.0255)} = 1.9$$

Entering the values of ISE and NE into our model and resimulating produces the result seen in Fig. 4.43. Adding ISE and NE significantly improves our model for low base-emitter voltages.

By this point in the extraction procedures, the simulated Gummel characteristics and measured Gummel characteristics appear to be very closely matched. Figure 4.44(*a*) illustrates just the collector current of the Gummel curve for both the measured values and the simulated values. The error between the two curves is 2.06 percent, as shown in Fig. 4.44(*b*). Comparing the base currents of the Gummel curves produces a similar result. At this point, our DC model of the bipolar transistor meets our criteria for a practical device model. A comparison of the measured and simulated family of curves is shown in Fig. 4.45.

```
* * * * * * * * * * LINREG Results * * * * * * * *
*
* For the equation Y = a + b*X
*
* a    =    -0.27216E+02
* b    =     0.20600E+02
* chi2 = 0.000
*
```

Figure 4.42 LINREG output used to extract ISE and NE.

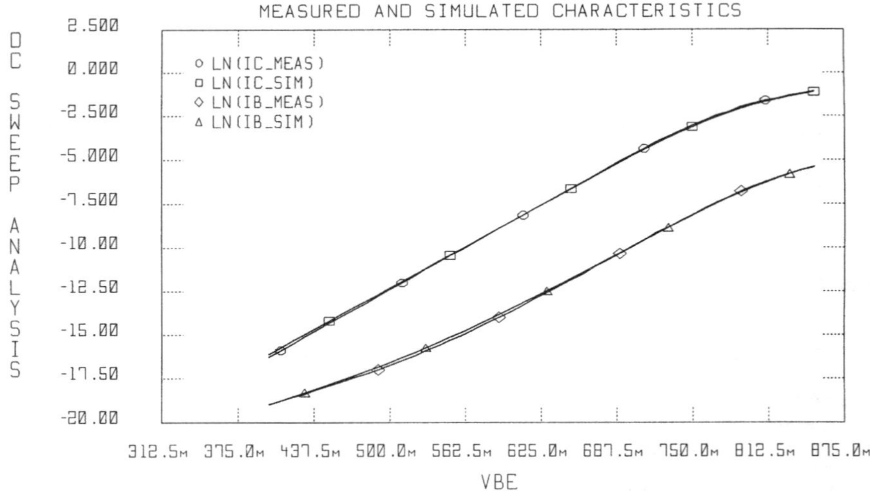

Figure 4.43 Comparison of measured and simulated model.

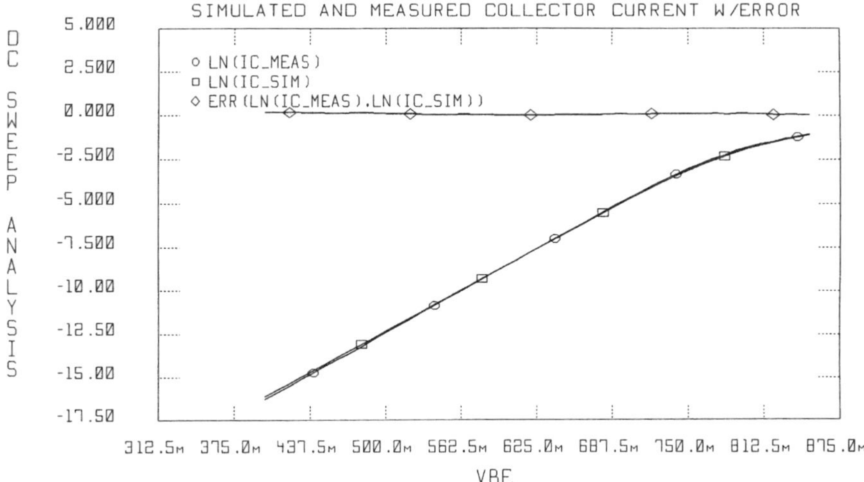

Figure 4.44(*a*) Measuring error between measured and simulated collector current curves.

The Gummel parameters, IS, NF, BF, RC, RE, RB, IKF, ISE, and NE, characterize the DC behavior of the bipolar transistor under forward operation. If the transistor was to be operated in reverse mode, the corresponding reverse parameters, NR, BR, IKR, ISC, and NC, would replace the forward parameters. The reverse parameters are determined by reversing the terminal voltages and reproducing the Gummel plot. If the device is to be operated in both regions of operation, both the forward and reverse Gummel plots should be constructed, and both the forward and reverse parameters should be determined.

While the Gummel parameters describe the transistor under low, medium, and high V_{be} voltages, all of the Gummel data is taken with the collector-base junction at zero volts. For some transistors, reverse-biasing the base-collector junction produces an increase in the output current. This increase is known as base-width modulation or the Early voltage effect. In SPICE, the early voltage is modeled with the parameter VAF. As shown in Fig. 4.14, extracting VAF is done with the family of curves. VAF represents the voltage on the V_{ce} axis where a linear approximation to the family of curves would intersect. However, looking back at Fig. 4.26, our own transistor appears to have little if any slope to the output characteristics. For this reason, we will leave VAF at the default setting (by not specifying a value for VAF) of minus infinity. *Please note:* when you do specify a value for the early voltage, the VAF model parameter is always entered as a

Error report

$\ln(I_C$ meas)	$\ln(I_C$ sim)	Percent error
-16.2718E+00	-16.1081E+00	-1.0%
-15.8918E+00	-15.7288E+00	-1.0%
-15.5138E+00	-15.3498E+00	-1.1%
-15.1036E+00	-14.9707E+00	-0.9%
-14.7167E+00	-14.5916E+00	-0.9%
-14.3307E+00	-14.2126E+00	-0.8%
-13.9509E+00	-13.8335E+00	-0.8%
-13.5956E+00	-13.4543E+00	-1.0%
-13.1987E+00	-13.0755E+00	-0.9%
-12.7945E+00	-12.6964E+00	-0.8%
-12.3970E+00	-12.3172E+00	-0.6%
-12.0145E+00	-11.9383E+00	-0.6%
-11.6333E+00	-11.5593E+00	-0.6%
-11.2217E+00	-11.1800E+00	-0.4%
-10.8374E+00	-10.8014E+00	-0.3%
-10.4537E+00	-10.4223E+00	-0.3%
-10.0920E+00	-10.0432E+00	-0.5%
-96.9015E-01	-96.6431E-01	-0.3%
-93.0961E-01	-92.8550E-01	-0.3%
-89.0286E-01	-89.0654E-01	0.0%
-85.2422E-01	-85.2825E-01	0.0%
-81.5673E-01	-81.4978E-01	-0.1%
-77.6272E-01	-77.7169E-01	-0.1%
-73.8064E-01	-73.9389E-01	-0.2%
-69.9419E-01	-70.1679E-01	-0.3%
-66.1359E-01	-66.4079E-01	-0.4%
-62.3634E-01	-62.6590E-01	-0.5%
-58.7173E-01	-58.9252E-01	-0.4%
-54.6532E-01	-55.2196E-01	-1.0%
-50.8530E-01	-51.5508E-01	-1.4%
-47.2824E-01	-47.9319E-01	-1.4%
-43.7565E-01	-44.3796E-01	-1.4%
-39.9867E-01	-40.9175E-01	-2.3%
-36.6360E-01	-37.5716E-01	-2.6%
-33.4161E-01	-34.3610E-01	-2.8%
-30.4031E-01	-31.3109E-01	-3.0%
-27.6098E-01	-28.4439E-01	-3.0%
-24.8099E-01	-25.7702E-01	-3.9%
-22.7303E-01	-23.2996E-01	-2.5%
-19.8486E-01	-21.0291E-01	-5.9%
-18.0363E-01	-18.9579E-01	-5.1%
-16.2760E-01	-17.0815E-01	-4.9%
-15.0058E-01	-15.3805E-01	-2.5%
-13.2126E-01	-13.8510E-01	-4.8%
-12.2963E-01	-12.4688E-01	-1.4%
-10.9602E-01	-11.2239E-01	-2.4%

RMS Error = 2.09%

Figure 4.44(*b*) Error between simulated and measured curves.

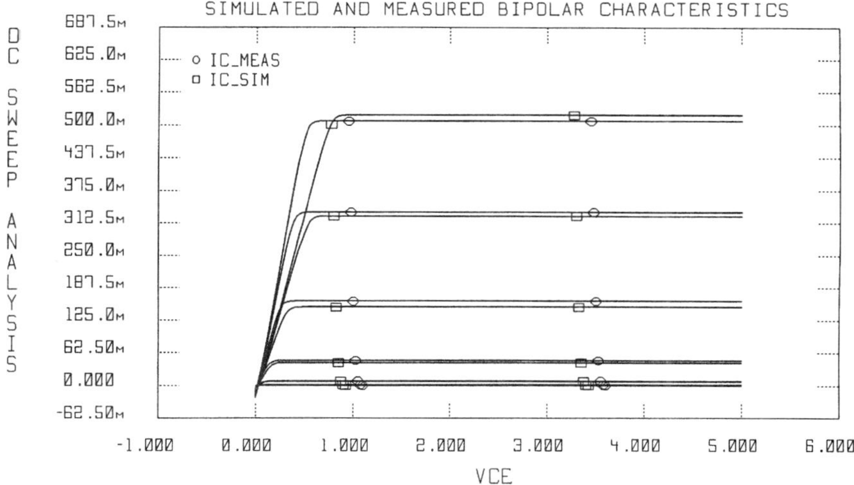

Figure 4.45 Comparison between simulated and measured family of curves.

positive number. Again, if the transistor was to be operated in the reverse mode, you would specify VAR rather than VAF.

With the DC parameters out of the way, we can turn our attention to the capacitance model parameters. The capacitance parameters shown in Table 4.2 are identical in function to the capacitance parameters of the diode. CJE, VJE, and MJE are the base-emitter equivalent to CJO, VJ, and M for the diode. CJC, VJC, and MJC are the capacitance parameters of the base-collector junction, and for integrated circuits, CJS, VJS, and MJS are the capacitance parameters of the collector-substrate junction.

To extract each of these parameters, we must measure the junction capacitance while changing the reverse-bias potential of the junction. The first measurement was performed on the base-collector junction of the transistor. Figure 4.46 shows the test setup used for the measurements, and Table 4.8 illustrates the measured values.

From this data, we need to extract the parameters CJC, VJC, and MJC, and we will perform the extraction with the help of the CJ_CALC program introduced in earlier. (Appendix B discusses using the CJ_CALC program for those readers who need a refresher.) The capacitance data was entered into the CJ_CALC program, and the program produced values for CJC, VJC, and MJC. Table 4.9 illustrates the capacitance values calculated by CJ_CALC. Figure 4.47(*a*) illustrates the measured value of capacitance, and Fig. 4.47(*b*) shows a comparison between the measured and simulated characteristics. The error between the measured and simulated characteristics is less than 0.3 percent!

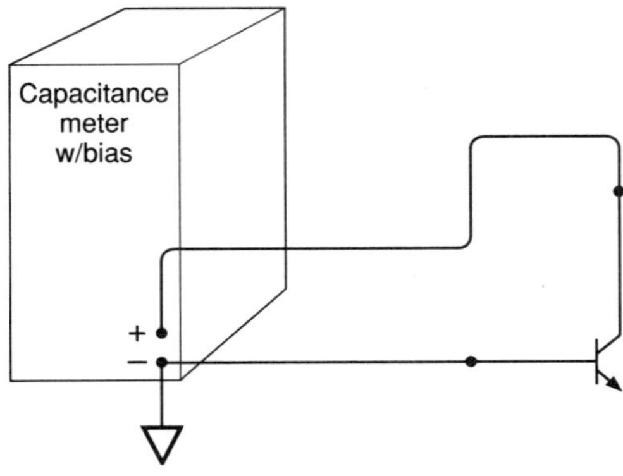

Figure 4.46 Measurement setup for base-collector capacitance extraction.

TABLE 4.8 Measured Base-Collector Junction Capacitance

Measured capacitance	V_{bc}
1.152E-11	−0.000E-01
1.088E-11	−1.000E-01
1.037E-11	−2.000E-01
9.934E-12	−3.000E-01
9.550E-12	−4.000E-01
9.213E-12	−5.000E-01
8.937E-12	−6.000E-01
8.674E-12	−7.000E-01
8.429E-12	−8.000E-01
8.212E-12	−9.000E-01
8.030E-12	−1.000E+00
6.686E-12	−2.000E+00
5.927E-12	−3.000E+00
5.394E-12	−4.000E+00
5.019E-12	−5.000E+00
4.707E-12	−6.000E+00
4.473E-12	−7.000E+00
4.258E-12	−8.000E+00
4.095E-12	−9.000E+00
3.928E-12	−1.000e+01

TABLE 4.9 Base-Collector Junction Capacitance Parameters

Parameter name	Value
CJC	11.5 pf
VJC	0.63
MJC	0.38

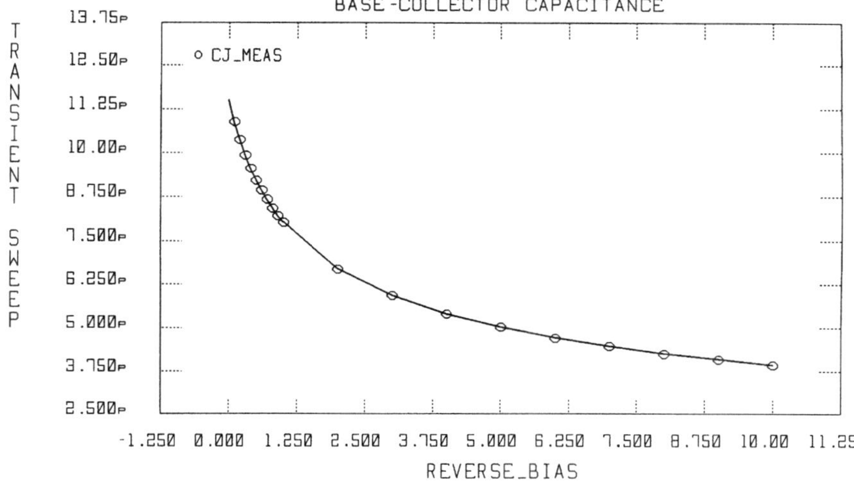

Figure 4.47(*a*) Measured base-collector junction capacitance.

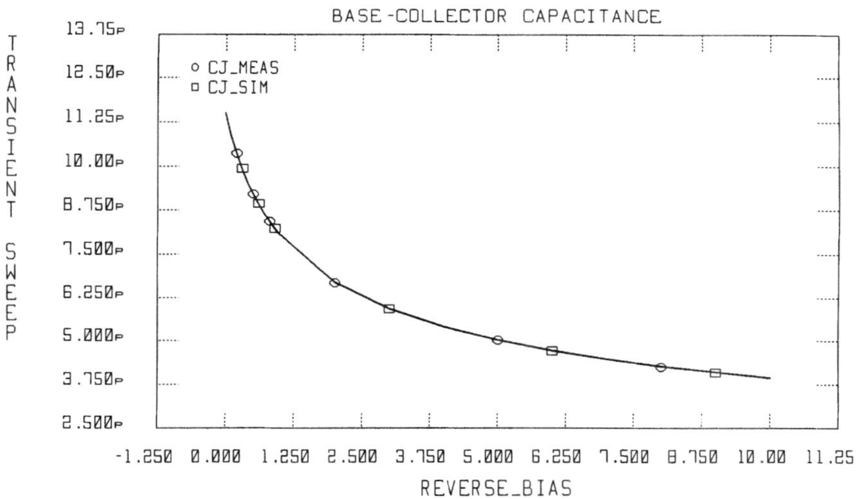

Figure 4.47(*b*) Measured and simulated base-collector junction capacitance.

The capacitance test leads were switched to the base-emitter junction and the capacitance voltage characteristics were measured. The capacitance values were entered into the CJ_CALC program and values of CJE, VJE, and MJE were calculated. Table 4.10 illustrates the capacitance parameters calculated for the base-emitter junction. Figure 4.48(*a*) illustrates the measured capacitance, and Fig. 4.48(*b*) compares the simulated characteristic with the measured values. The

TABLE 4.10 Base-Emitter Junction Capacitance Parameters

Parameter name	Value
CJE	19.0 pf
VJE	0.87
MJE	0.48

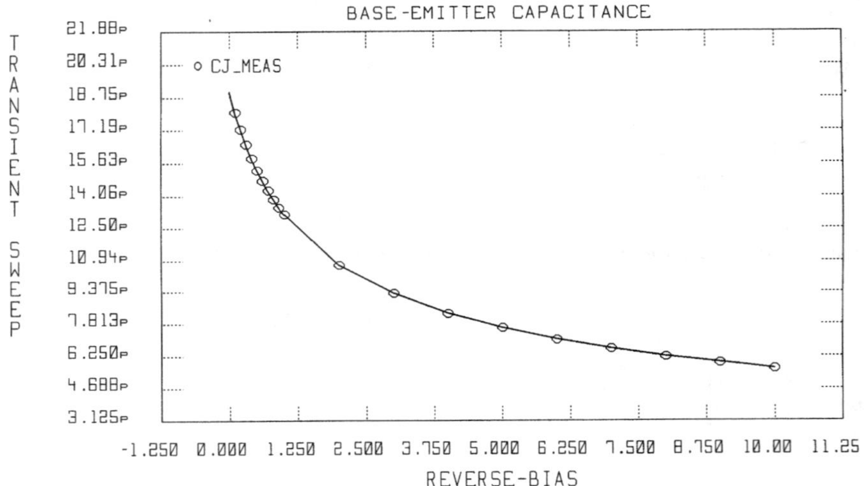

Figure 4.48(*a*) Measured base-emitter junction capacitance.

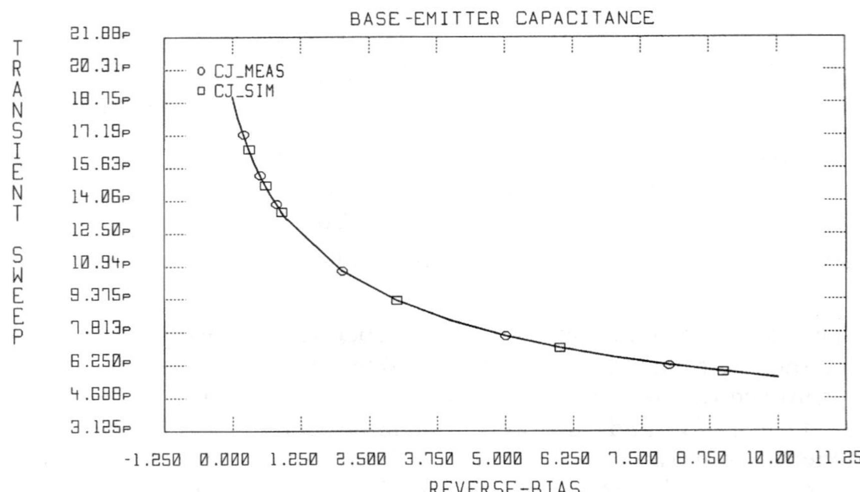

Figure 4.48(*b*) Measured and simulated base-emitter junction capacitance.

error between the measured and simulated characteristics is again less than 0.3 percent!

Since our transistor is a discrete device, there is no substrate connection. Because of this, the substrate capacitance parameters CJS, MJS, and VJS will be left at their default of zero. FC is the capacitance factor which was originally discussed in the diode chapter. As we did with the diode, FC will be left at the default of 0.5 for our bipolar model.

The final capacitance parameter of interest is TF. TF is determined from the unity-gain frequency (f_T). The unity-gain frequency can often be found on the data sheet. From our data sheet for this device, the unity-gain frequency is shown as 300 MHz. Once the values of RC and CJC have been determined, TF can be calculated from Eq. 4.11. Our calculation for TF is shown in Eq. 4.21. From this calculation, TF for our model will be set to 518 pS.

$$\mathrm{TF} = \frac{1}{(2 * \pi * (300\ \mathrm{MHz}))} - (11.5\ \mathrm{pf})*(1.4\ \Omega) = 514.4\ \mathrm{pS} \quad (4.21)$$

Table 4.11 is a list of the model parameters we calculated for our model. Calculating parameters for the bipolar transistor is essential for an accurate simulation model. It is unfortunate that, even today, many of our bipolar manufacturers fail to realize the importance of including plots of the Gummel characteristics and the capacitance-

TABLE 4.11 Calculated Model Parameters for the Motorola 2N2222a Bipolar Transistor

Parameter name	Value
IS	26.28 fa
NF	1.02
BF	188
IKF	0.2a
RC	1.1 Ω
RE	0.026 Ω
RB	4.5 Ω
ISE	1.5 pa
NE	1.9
CJC	11.5 pf
VJC	0.63
MJC	0.38
CJE	19 pf
VJE	0.87
MJE	0.48
TF	514 pS

vs.-voltage characteristics of the transistor on the device data sheet. As more and more designers turn to analog simulation to ensure the quality of their design, and as device manufacturers recognize the sales benefit of including data sheets of more value to the designer, better data sheets will become available. You can already see this trend beginning, and many device manufacturers already provide SPICE models of their components. But until all manufacturers provide models, understanding how to extract the parameters from measured devices will be crucial to circuit simulation.

References

1. Giuseppe Massobrio and Paolo Antognetti, *Semiconductor Device Modeling with SPICE,* McGraw-Hill, New York, 1988, 1993.
2. Andrei Vladimirescu, *The Spice Book,* John Wiley, New York, 1994.
3. H. K. Gummel, "A Charge Control Relation for Bipolar Transistors," *Bell Systems Technical Journal,* vol. 49, January 1970, pp. 115–120.
4. Ian Getreu, *Modeling the Bipolar Transistor,* Tektronix, Inc., part #062-2841-00.

Chapter

5

Modeling the Junction Field Effect Transistor

Introduction

In this chapter, you will learn to model the junction field-effect transistor (JFET). The JFET used in SPICE is modeled with the equivalent circuit shown in Fig. 5.1, and, as such, many of the model parameters and extraction techniques you learned in Chap. 3 will be applied to the gate-drain and gate-source diodes. For these junctions, you will be extracting parameters such as the reverse-leakage current (IS), the zero-bias capacitance parameters (CGD) and (CGS), and the built-

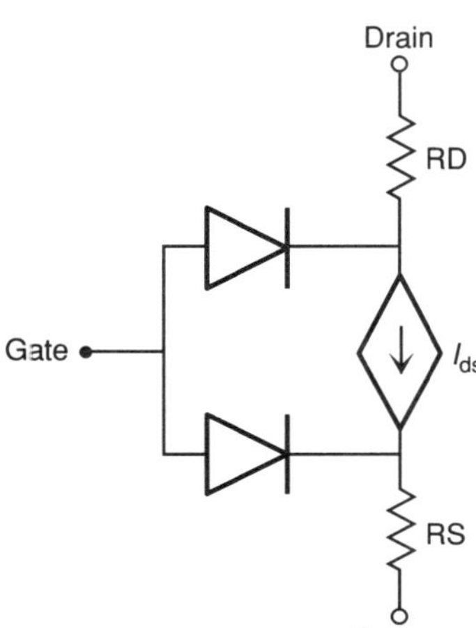

Figure 5.1 SPICE JFET equivalent circuit.

in potential (PB). There are 12 JFET model parameters; of those 12, 4 are basic diode parameters. Again, understanding the diode parameters is essential to understanding the JFET transistor model.

If you have already read Chap. 3 of this text, dealing with modeling junction diodes, you are now ready to model a junction field-effect transistor. If you skipped Chap. 3 and turned directly to this section, you should first read the section on modeling the diode. The information learned about the diode parameters and the modeling and extraction techniques covered in the diode chapter will be used in this chapter and others. The diode chapter also introduced the programs CJ_CALC and LINREG. We will make extensive use of these programs throughout this chapter, too.

Throughout this chapter, various circuits and models will be described. All of the circuits and models are found in the CH5 subdirectory of the disk bound to this text. The utility programs CJ_CALC, CJB, LINREG, and LRB are found in the UTIL subdirectory of the disk. The disk must be installed on your hard disk and uncompressed before you can access these files.

As we have throughout this text, we will focus on building *practical* JFET models in this chapter. But because the JFET model has the fewest number of model parameters and the least flexibility in fitting the model to measured data, some JFETs cannot be modeled accurately in all regions of operation. This fact was recognized by Curtice[1] who modified the original JFET equations proposed by Shichman[2]. But our goal is still to obtain a model which is accurate to within 5 percent rms error.

We will begin by looking at the standard SPICE JFET model parameters, defining and describing each in detail and explaining how each one is extracted. Following the explanation, a real junction field-effect transistor will be modeled.

Modeling Approach

In keeping with our philosophy of practical device modeling, typical junction field-effect transistor data sheets will be inspected for information which might be useful in developing device models. What cannot be obtained from the data sheet will be taken from measured I-V and C-V characteristics. After the initial set of model parameters has been chosen, simulations will be run on the model to optimize the final parameter values and to estimate the accuracy of our model.

JFET Model Parameters

Again, an introduction to the model parameters is necessary. Table 5.1 lists the 12 junction field-effect transistor model parameters. Also

TABLE 5.1 Junction Field-Effect Model Parameters

Name	Parameter	Units	Default
VTO	Zero-bias pinch-off voltage	V	0.0
BETA	Transconductance	A/V^2	1.0E-4
LAMBDA	Channel-length modulation	1/V	0.0
RD	Drain ohmic resistance	Ohm	0.0
RS	Source ohmic resistance	Ohm	0.0
IS	Gate junction leakage current	A	1.0E-14
CGD	Zero-bias G-D junction cap.	F	0.0
CGS	Zero-bias G-S junction cap.	F	0.0
PB (VJ)	Bulk junction potential	V	0.8
FC	Coefficient for cap formula	—	0.5
	Noise Parameters		
KF	Flicker-noise coefficient	—	0.0
AF	Flicker-noise exponent	—	1.0

shown in Table 5.1, along with the name, is the default value of each parameter. Every model parameter has a default value assigned inside the program. The parameter values found on the .MODEL statements simply override the default values assigned within the program.

The first six parameters, VTO, BETA, LAMBDA, RD, RS, and IS, describe the DC I-V characteristics of the JFET. The next four parameters, CGD, CGS, PB (VJ) and FC, describe the capacitive characteristics of the model. The parameters KF and AF model the flicker-noise of the device (KF and AF are used during only an AC noise analysis).

These 12 model parameters completely characterize the junction field-effect transistor in SPICE. The parameters can be used to model both discrete and integrated-circuit transistors.

The DC JFET parameters

The most basic and most important parameters of the JFET model are VTO, BETA, and LAMBDA. VTO represents the pinch-off voltage of the JFET. This parameter is sometimes known as $V_{GS(off)}$ on the data sheet. VTO determines the gate voltage at which the device enters the cutoff region of operation and the device current effectively goes to zero. Beta is the transconductance of the device. To better understand these parameters, the equations used to describe the drain current of the JFET are shown in Eqs. 5.1, 5.2, and 5.3. Figure 5.2 illustrates the two regions of operation described by Eqs. 5.1 and 5.2.

$$I_{ds} = 2 * \text{BETA} * \left(\frac{V_{gs} - \text{VTO} - V_{ds}}{2} \right) * V_{ds} * (1 + \text{LAMBDA}*V_{ds}) \quad (5.1)$$

for linear region $|V_{gs} - \text{VTO}| > V_{ds}$ and $0 < |V_{gs}| < |\text{VTO}|$

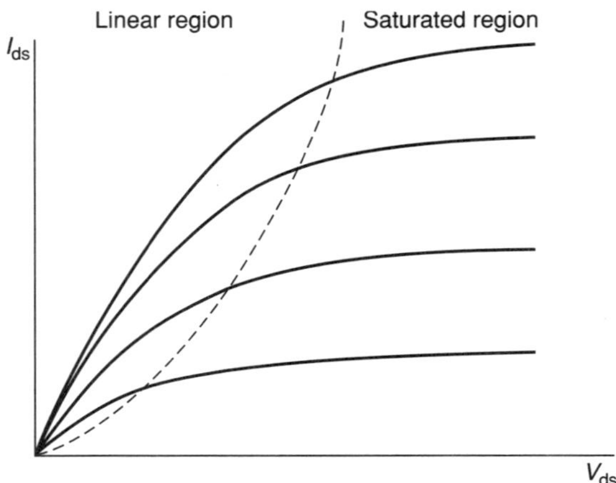

Figure 5.2 Linear and saturated regions of operation for a JFET.

$$I_{ds} = \text{BETA} * (V_{gs} - \text{VTO})^2 * (1 + \text{LAMBDA}*V_{ds}) \tag{5.2}$$

for saturated region $V_{ds} >= |V_{gs} - \text{VTO}|$
and $0 < |V_{gs}| < |\text{VTO}|$

$$I_{ds} = 0 \tag{5.3}$$

for cutoff region $|V_{gs}| > |\text{VTO}|$

From Eqs. 5.1 and 5.2, BETA represents the proportionality constant between the drain current and the terminal voltages. VTO represents the voltage at which the transistor enters the cutoff region and stops conducting current. Both of these parameters are determined from the I-V characteristics of the transistor.

The I-V characteristic familiar to most designers is the family of curves of a JFET. A typical family of curves is shown in Fig. 5.3. Unfortunately, the family of curves does not provide us with enough information to determine BETA and VTO. To extract these two parameters, we need a different set of I-V characteristics. Figure 5.4(*a*) illustrates another set of I-V characteristics from the JFET. These characteristics are known as the *linear* characteristics of a JFET. The linear characteristics get their name from the fact that to obtain this curve, the transistor must remain biased in the linear region of operation. Figure 5.4(*b*) illustrates the measurement setup for the linear characteristics. From the linear characteristics, we will obtain VTO, BETA, RS, and RD.

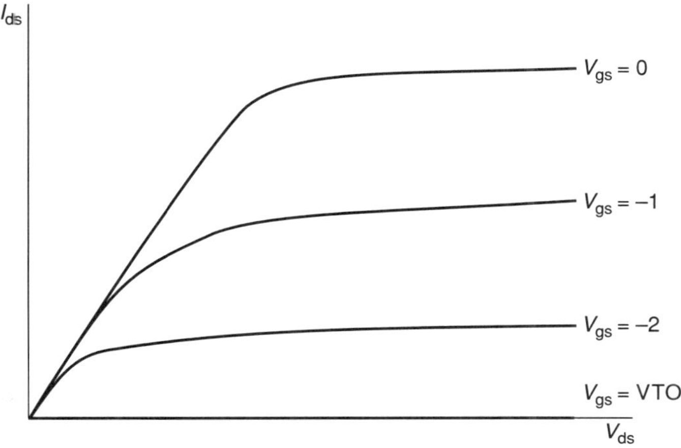

Figure 5.3 JFET family of curves.

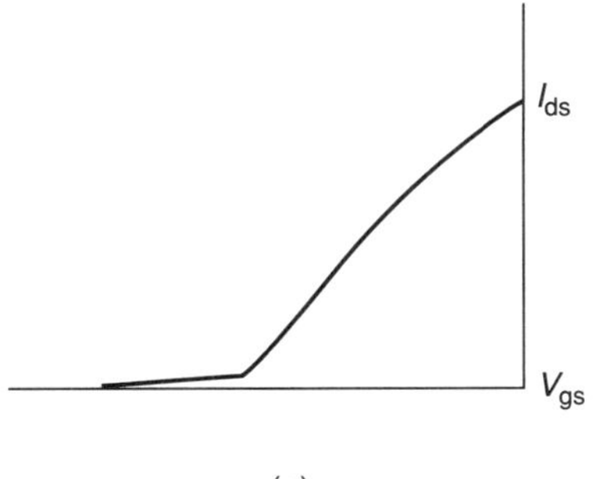

Figure 5.4(*a*) Linear JFET characteristics.

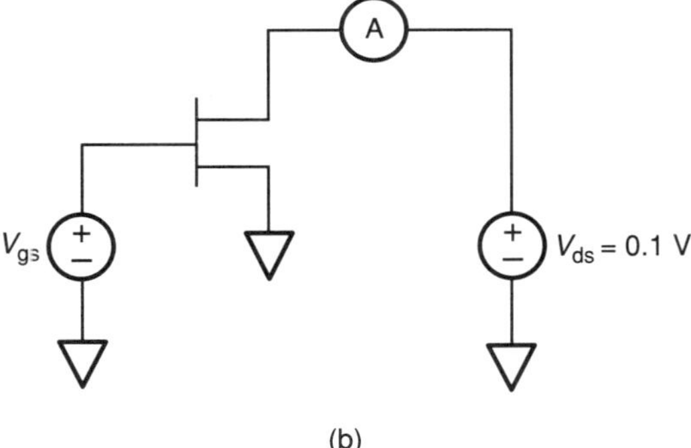

Figure 5.4(*b*) Measurement setup for linear characteristics.

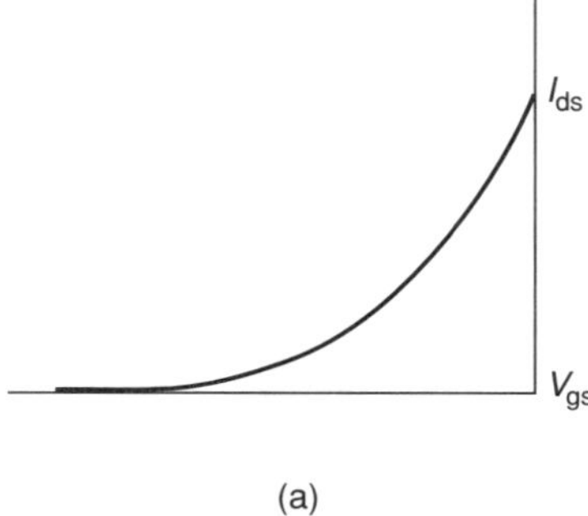

Figure 5.5(*a*) Saturated JFET characteristics.

Figure 5.5(*a*) illustrates another I-V characteristic, similar in appearance to the linear characteristics, the saturated characteristics. [While Figs. 5.4(*a*) and 5.5(*a*) are similar in appearance. Fig. 5.5(*a*) has a parabolic shape where Fig. 5.4(*a*) has a more linear shape. The difference in the graphs can be seen if you hold the book at an acute angle, with the bottom of the book closest to you and the top of the book farthest from you. In this position, you can see the curvature of Fig. 5.5(*a*).] The difference between the linear and saturated characteristics is found in the measurement setup shown in Fig. 5.5(*b*). The measurement setup in Fig. 5.4(*b*) holds V_{ds} to a small voltage value (100 mV). Holding V_{ds} below $|V_{gs}|$ forces the transistor to operate in the linear region only. The measurement setup in Fig. 5.5(*b*) holds V_{ds} at 5 V. By keeping V_{ds} greater than $|V_{gs}|$, the transistor is forced to operate in the saturated region of operation. The drain current in Figs. 5.4(*a*) and 5.5(*a*) are the drain currents described by Eqs. 5.1 and 5.2, respectively. From the saturated characteristics, we can determine a value for VTO and BETA.

But a question comes to mind: Which of these two tests should we use to extract the parameters VTO and BETA? You should use the

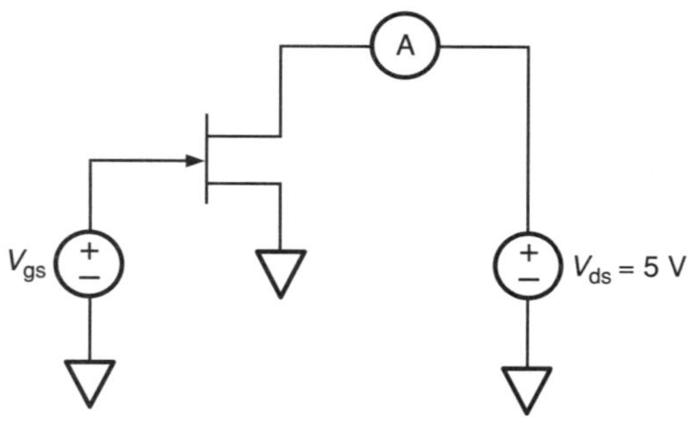

Figure 5.5(*b*) Measurement setup for saturated characteristics.

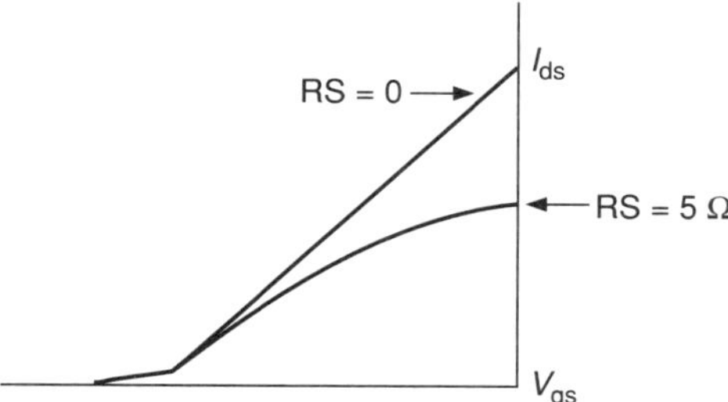

Figure 5.6 Linear characteristics with and without source series resistance.

linear test whenever you can, because only the linear test allows you to determine RS (and usually RD) in addition to BETA and VTO. As illustrated in Fig. 5.6, RS and RD have a pronounced influence on the I-V relationship of the JFET. Since the saturated test offers no means to determine RS and RD, *the value of BETA calculated in the saturated test is usually inaccurate.*

This logically leads to another question: If the linear characteristics are the preferred way to determine VTO and BETA, why even bring up the saturated characteristics? The reason for the discussion of the saturated characteristics is twofold: (1) often, the data sheet of a JFET will contain a plot of the saturated characteristics. For many users, the data sheet characteristics may be the only data available to determine BETA and VTO; and (2) because it is common for the saturated characteristics to appear on the data sheet, readers might confuse the two characteristics and try to apply the linear extraction techniques to the saturated data. The results of applying the linear extraction techniques to the saturated characteristics are erroneous results for BETA, RS, and RD, as well as one frustrated SPICE user. So, for these reasons, extraction procedures for both of these characteristics will be presented.

But before we go further, three programs have been created which help you simulate the linear, saturated, and family of curves of a JFET. The first of these programs is called LIN.BAT. LIN.BAT is a batch procedure which runs RSPICE on a file named LIN.CIR. The LIN.CIR file is a SPICE input file which simulates the linear characteristics of a JFET. The LIN.CIR file reads the JFET model parameters from a file named NCHAN.MOD. The output from the LIN.CIR simulation is shown in Fig. 5.7. The program SAT.BAT is similar to LIN.BAT but generates the saturated characteristics of the JFET. The output of SAT.BAT is shown in Fig. 5.8. The final program, FAMILY.BAT simulates the family of curves of a JFET. The output of FAMILY.BAT is shown in Fig. 5.9. All three of these programs are stored in the `RCGV34\CH5` subdirectory. To run these programs, change into the CH5 subdirectory and at the DOS prompt type:

```
LIN
```

or

```
SAT
```

or

```
FAMILY
```

The resulting RGRAPH plots should match Figs. 5.7, 5.8, and 5.9.

We will begin our discussion by illustrating techniques to determine VTO and BETA from the saturate characteristics of a JFET. But remember, you should use the saturated characteristics only when the linear characteristics are unavailable. To extract VTO from the saturated characteristics, the I-V characteristic so often found on the transistor data sheet, we must determine the V_{gs} voltage where the transistor begins to conduct current. Since the transistor is biased in the saturated region of operation, Eq. 5.2 governs the drain current. If we set I_{ds} to zero, Eq. 5.2 reduces to Eq. 5.4.

$$0 = (V_{gs} - \text{VTO})^2 * (1 + \text{LAMBDA} * V_{ds}) \tag{5.4}$$

Since LAMBDA is usually small (LAMBDA has a default value of 0), the second term in Eq. 5.4 can be eliminated, resulting in Eq. 5.5.

$$0 = (V_{gs} - \text{VTO})^2 \tag{5.5}$$

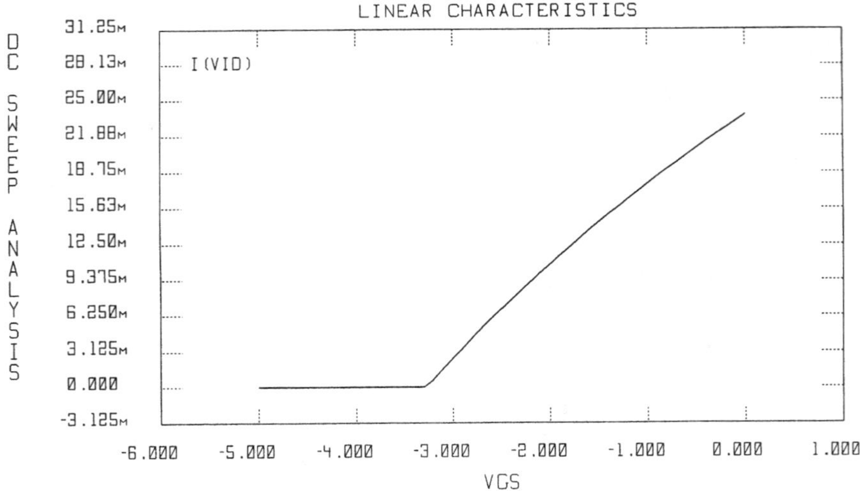

Figure 5.7 LIN.BAT simulation result.

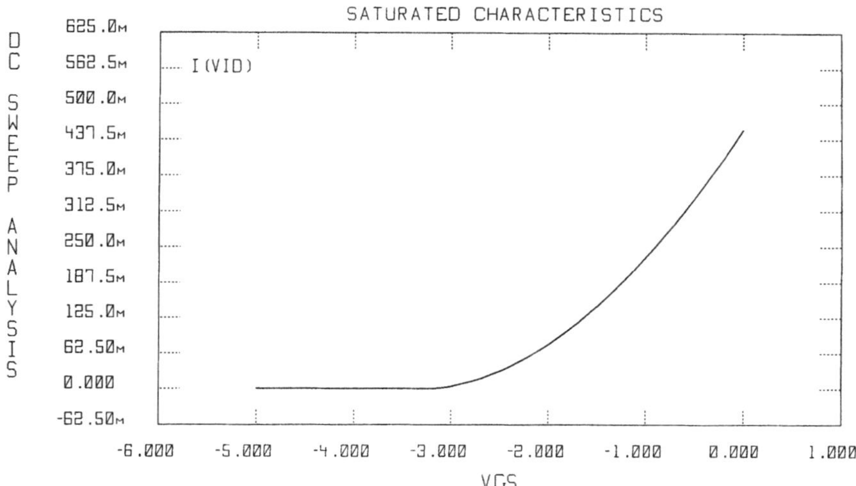

Figure 5.8 SAT.BAT simulation result.

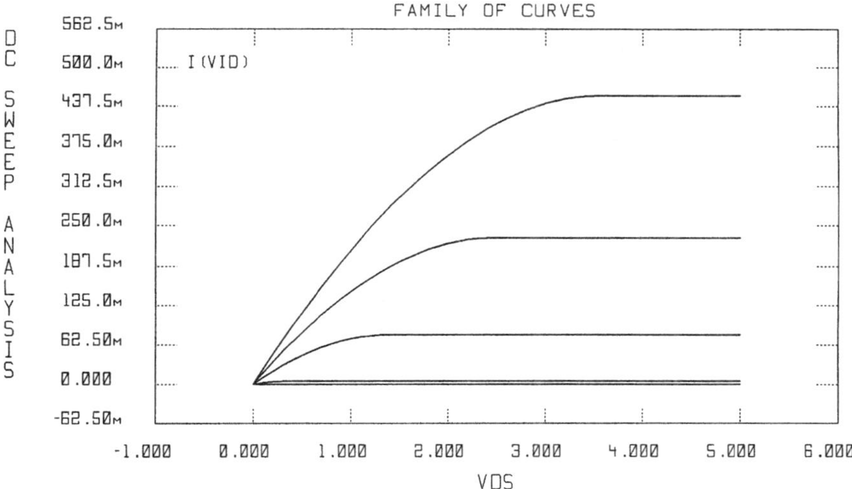

Figure 5.9 FAMILY.BAT simulation result.

Taking the square root of both sides leaves Eq. 5.6.

$$0 = V_{gs} - \mathrm{VTO} \tag{5.6}$$

or

$$V_{gs} = \mathrm{VTO}\ \big|_{\text{at Id} = 0} \tag{5.7}$$

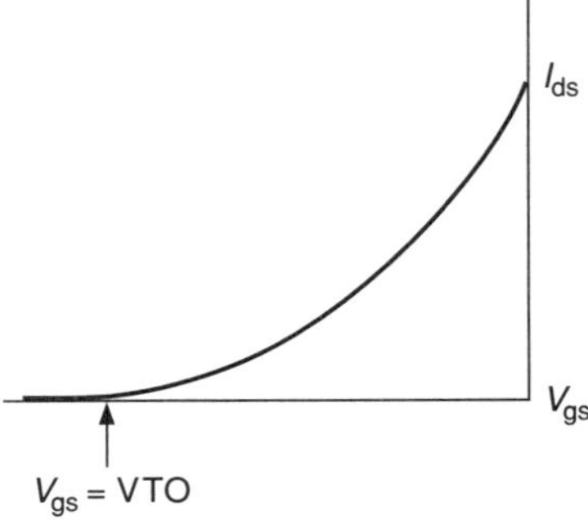

Figure 5.10 Extracting VTO from the saturated JFET characteristics.

Thus, VTO can be found by locating the gate voltage (V_{gs}) where the transistor starts conducting current. An example of this is shown in Fig. 5.10.

Once VTO has been determined, we can begin to extract our other DC parameters. The next parameter we need to determine is LAMBDA. LAMBDA determines the slope of the output characteristics shown in the family of curves. LAMBDA is extracted by extrapolating the slope of the output characteristics onto the V_{ds} axis. The extraction is shown in Fig. 5.11(*a*). *Note:* Many transistors have virtually flat output characteristics. This is modeled with LAMBDA set to zero. Zero is the default setting of LAMBDA. If your transistor characteristics display no observable slope to the characteristics, leave LAMBDA at zero.

While some data sheets include a plot of the family of curves, most will not. In such a case, the family of curves can be obtained with a curve tracer or parametric analyzer such as the HP 4142a. The measurement setup for the family of curves is shown in Fig. 5.11(*b*).

An alternative to extrapolating the output characteristics' slope is to use a simple regression program and determine the slope and

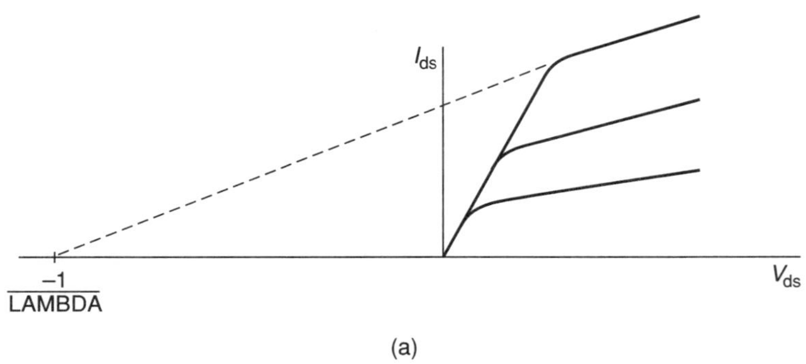

Figure 5.11(*a*) Extracting LAMBDA from the family of curves.

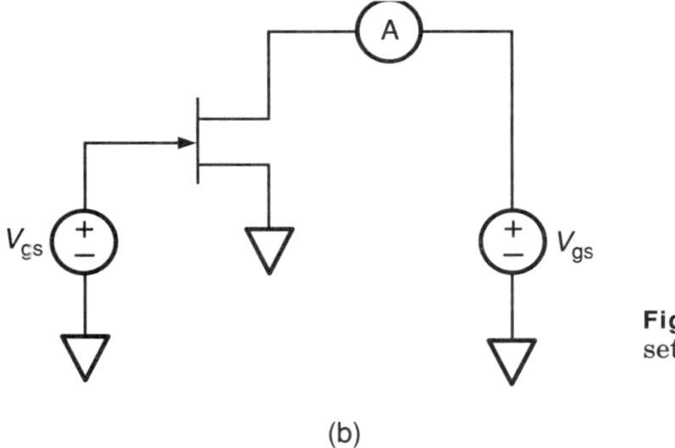

Figure 5.11(*b*) Measurement setup for family of curves.

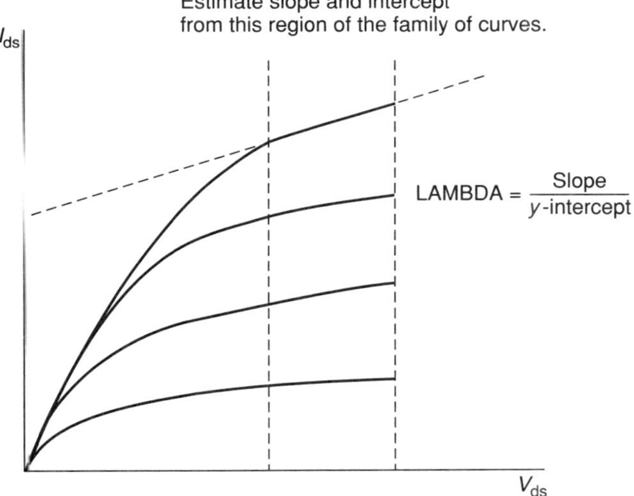

Figure 5.12 Extracting LAMBDA from slope of family of curves.

intercept of the output characteristics as shown in Fig. 5.12. The LINREG program, introduced in Chap. 3, is a regression program which will calculate the slope and intercept of a line. With the slope and intercept determined, LAMBDA may be calculated from Eq. 5.8.

$$\text{LAMBDA} = \frac{\text{slope}}{y\text{-intercept}} \tag{5.8}$$

With LAMBDA determined, we can return to the saturated I-V characteristic to determine BETA. With a plot of I_{ds} vs. V_{gs}, measure the highest value of I_{ds}, and determine the value of V_{gs} which corresponds to this value of I_{ds}. With the values of VTO and LAMBDA determined, rearranging Eq. 5.2 yields Eq. 5.9. BETA can be deter-

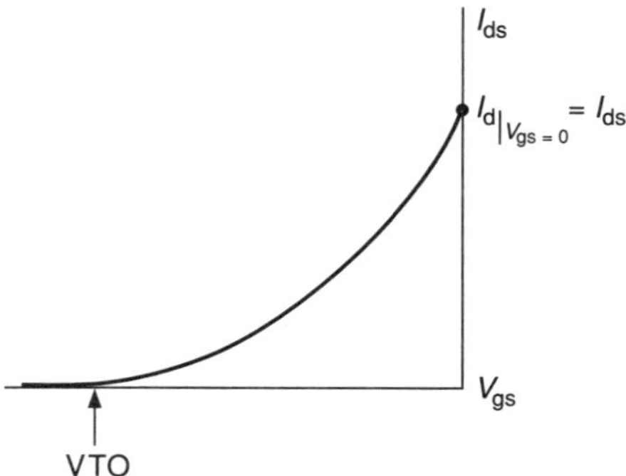

Figure 5.13 Measuring the maximum drain current.

mined from entering VTO, LAMBDA, and a value of I_{ds} and V_{gs} measured from the saturated curve. Usually the value of I_{ds} should be as high as possible. (V_{gs} should be close or equal to zero.) This measurement is illustrated in Fig. 5.13.

$$\text{BETA} = \frac{((V_{gs} - \text{VTO})^2 * (1 + \text{LAMBDA} * V_{ds}))}{I_{ds}} \tag{5.9}$$

If the drain current is measured with the gate voltage set to zero (this current is also known as I_{dss}), Eq. 5.9 reduces to Eq. 5.10.

$$\text{BETA} = \frac{(\text{VTO}^2 * (1 + \text{LAMBDA} * V_{ds}))}{I_{dss}} \tag{5.10}$$

If LAMBDA is a small, or the slope of the output characteristics is too slight to measure, Eq. 5.9 can be reduced to Eq. 5.11.

$$\text{BETA} = \frac{(V_{gs} - \text{VTO})^2}{I_{dss}} \tag{5.11}$$

Or, at $V_{gs} = 0$:

$$\text{BETA} = \frac{\text{VTO}^2}{I_{dss}} \tag{5.12}$$

Equation 5.12 is a fast, easy means to calculate BETA. *But the value of BETA determined with Eq. 5.12 will be inaccurate because we have neglected the effects of LAMBDA and the series resistances (RD and RS).*

The preferred method to determine VTO and BETA is with the linear I-V characteristics. The reason the linear characteristics are preferred is that the linear extraction method gives us a way to determine the source resistance RS and the drain resistance RD. Neither

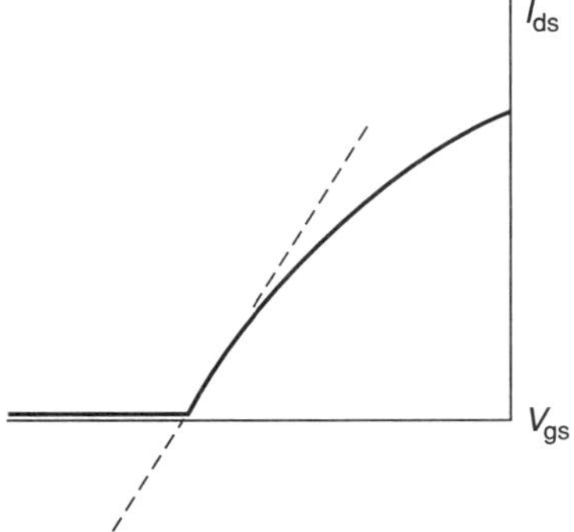

Figure 5.14 Straight-line approximation to linear characteristics at low drain currents.

of these resistance values can be determined from the saturated characteristics.

To determine VTO and BETA from the linear characteristics of the JFET, make a straight-line approximation to the characteristics at low drain currents, as shown in Fig. 5.14. The intersection of this line on the V_{gs} axis is used to calculate VTO, the slope of the line is used to calculate BETA, and the intersection of the line on the y-axis is used to determine RD and RS. These extractions are illustrated in Fig. 5.15. VTO is found from the intersection with the x-axis and Eq. 5.13. BETA is calculated from the slope of the line and Eq. 5.14.

$$\text{VTO} = V_{\text{gs(intercept)}} - \frac{V_{\text{ds}}}{2} \tag{5.13}$$

$$\text{BETA} = \frac{\text{slope}}{(2 * V_{\text{ds}})} \tag{5.14}$$

To determine RS and RD, we must measure the drain current at $V_{gs} = 0$. This current is known as I_{dss}, but we shall label the current $I_{d(low)}$.

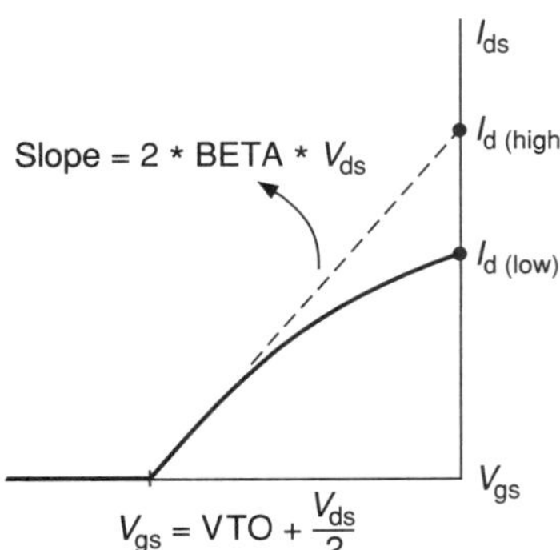

Figure 5.15 Linear JFET characteristics.

We also need to know the drain current where the straight-line approximation intersects the curve. This current is known as $I_{d(high)}$. Using these two currents, we can determine the drain and source resistance values.

$$I_{d(high)} - I_{d(low)} = 4*BETA*I_{d(low)}*RS* \left(-V_{ds} - V_{gs} - I_{d(low)}*RS + VTO + \frac{V_{ds}}{2}\right) \tag{5.15}$$

Finding the source resistance involves solving Eq. 5.15, which is a quadratic, in terms of RS. To ease this calculation, a simple FORTRAN program has been created which solves Eq. 5.15 for RS. The program is called RS_CALC and is stored in the UTIL subdirectory of the disk which accompanies this text. The RS_CALC program asks the user for the terminal voltages of the device, the calculated VTO and BETA, and the $I_{d(high)}$ and $I_{d(low})$ currents. The program solves Eq. 5.15 and determines the correct RS value.

Learning to use the RS_CALC program is beyond the scope of this chapter. The interested reader is directed to App. E for more information on using RS_CALC.

Since most JFETs are fabricated symmetrically and have a drain resistance equal to the source resistance, once RS has been determined, RD may be set equal to RS.

The last of the DC parameters is IS. In the diode model, IS represented the reverse leakage current of the diode. For the JFET, IS represents the reverse leakage between the gate and drain and between the gate and source. This parameter is often found on the data sheet of a JFET under the name *reverse gate leakage* (I_{GSS}). IS has very little influence on the I-V characteristics of the transistor unless you forward-bias the gate (which is almost never done with JFETs). Because IS has so little influence on the normal operation of the JFET, a practical value of IS can usually be obtained from the data sheet value for I_{GSS}. For those of you who need a more accurate value for the gate leakage, IS can be determined with the same test used to determine IS for the junction diode. The measurement setup to determine IS is shown in Fig. 5.16(*a*) and the actual extraction procedure is shown in Fig. 5.16(*b*).

The capacitive JFET parameters

With BETA, VTO, LAMBDA, RS, RD, and IS extracted, we can turn our attention to the capacitance parameters, CGS, CGD, PB (VJ), and FC. CGS is the zero-bias gate-source capacitance and is equivalent to the data sheet capacitance named C_{iss}. CGD is the zero-bias gate-

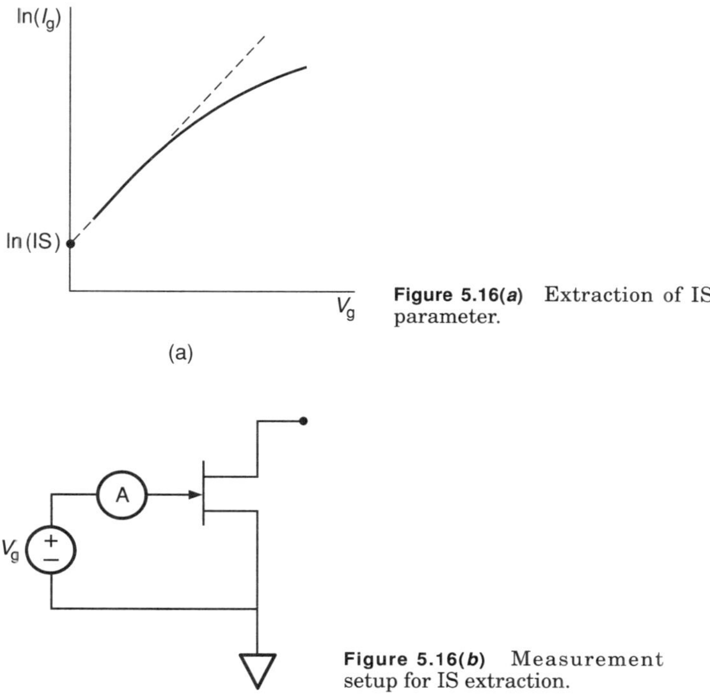

Figure 5.16(*a*) Extraction of IS parameter.

Figure 5.16(*b*) Measurement setup for IS extraction.

drain capacitance and is equivalent to the data sheet capacitance named C_{iss}. Both of these capacitance values are similar to CJO for the diode model.

While the data sheet often describes the value of C_{iss} and C_{rss}, the values stated on the data sheet are usually measured with a nonzero bias applied to the drain of the transistor. The values of CGD and CGS must be determined with no external bias applied to the JFET.

To determine the zero-bias capacitance values of the JFET, *both capacitances must be measured at the same time.* The measurement for CGS and CGD is shown in Fig. 5.17. If we assume the transistor is fabricated symmetrically, once the zero-bias gate capacitance has been determined, CGS and CGD can each be assigned a value equal to half of the measured capacitance.

PB is similar in form to the VJ model parameter introduced in the diode and bjt chapters, but for JFETs and MOSFETs, the authors of SPICE gave this parameter a different name. For our models, PB and VJ are identical in function and purpose but have different names.

What is notably missing from the list of JFET capacitance model parameters is the grading coefficient M. In SPICE, the JFET capaci-

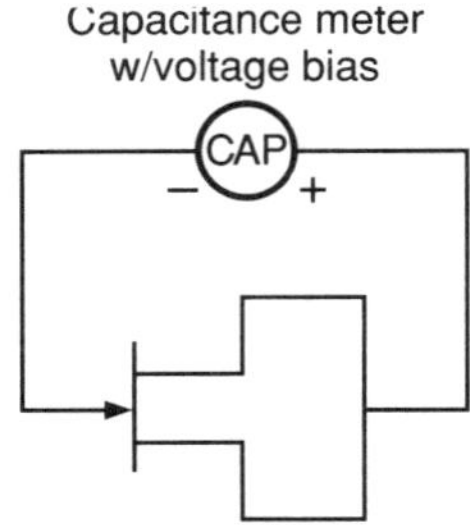

Figure 5.17 Measurement setup for CGD and CGS extraction.

tance model has set the value of M to 0.5, and this value cannot be changed by the user.

In the diode section, we learned to calculate VJ and M with the aid of the CJ_CALC program. We can still use the CJ_CALC program to determine the JFET capacitance parameter PB (set PB equal to the value CJ_CALC computes for VJ), but the value of M which is generated from CJ_CALC should be discarded. (*Note:* Many of the vendor-offered SPICE-like simulators available today have modified the M grading coefficient into a user-selectable model parameter. For these simulators, use the value of M which is generated from the CJ_CALC program.)

To determine PB for the JFET, we need to measure several different capacitance values as we change the reverse-bias voltage on the gate of the JFET. After entering the measured data into CJ_CALC, set PB equal to the VJ model parameter calculated by CJ_CALC. Refer to App. B for instructions on using the CJ_CALC program.

As with our diode and bipolar junction transistor model, FC will be left at the default value of 0.5.

JFET Model Extraction Demonstration

To demonstrate our modeling techniques, we will extract model parameters from a 2N4416 JFET. Figure 5.18 illustrates the data sheet information for this transistor.

While the data sheet information may be of use to the designer, only one piece of information appears to be immediately useful in device modeling: the reverse gate leakage current, I_{gss}. I_{gss} will be used as the value for the IS model parameter. The balance of the model parameters will be determined by I-V and C-V tests.

We will begin our modeling procedures by extracting VTO and BETA from the transistor. While the data sheet contains the saturated characteristics of the 2N4416 transistor, a more accurate value for VTO and BETA can be obtained from the linear characteristics. The linear characteristics of the JFET were measured with an HP 4142a

Electrical Characteristics (T_A = 25°C)

Symbol	Parameter	Conditions	Min	Typ	Max	Units
BV_{GSS}	Gate-Source Breakdown Voltage	V_{DS} = 0V, I_G = −1 μA	−25	−40		V
I_{DSS}	Zero Gate Voltage Drain Current	V_{DS} = 15V, V_{GS} = 0V	1.0	10	20	mA
g_{fs}	Forward Transconductance	V_{DS} = 15V, V_{GS} = 0V	3.0	5.5	7.0	mmhos
g_{fs}	Forward Transconductance	V_{DG} = 15V, I_D = 200 μA		1.1		mmhos
I_{GSS}	Reverse Gate Leakage	V_{GS} = −20V, V_{DS} = 0V		−5.0	−100	pA
$r_{DS(ON)}$	ON Resistance	V_{DS} = 100 mV, V_{GS} = 0V	100	175	500	Ω
$V_{GS(OFF)}$	Pinch Off Voltage	V_{DS} = 15V, I_D = 1 nA	−0.7	−3.5	−6.0	V
g_{os}	Output Conductance	V_{DG} = 15V, I_D = 1 mA, f = 1 kHz		10		μmhos
C_{rss}	Feedback Capacitance	V_{DG} = 15V, V_{GS} = 0V		0.7	0.9	pF
C_{iss}	Input Capacitance	V_{DS} = 15V, V_{GS} = 0V		3.5	4.0	pF
e_n	Noise Voltage	V_{DG} = 15V, I_D = 1 mA, f = 100 Hz		8.0		nV/√Hz
NF	Noise Figure	V_{DG} = 15V, I_D = 5 mA, R_G = 1 kΩ, f = 400 MHz		2.2	4.0	dB
G_{PS}	Power Gain	V_{DG} = 15V, I_D = 5 mA, f = 400 MHz		12		dB

Figure 5.18 Typical JFET data sheet values.

parametric analyzer. The measured characteristics are shown in Fig. 5.19.

VTO and BETA are determined from the slope and intercept of the low-I_{ds} portion of the curve, as shown in Fig. 5.20. The I-V values between $V_{gs} = -3.5$ and $V_{gs} = -3$ shown in Fig. 5.21 were entered into the LINREG program.

LINREG is a linear regression program which determines the slope and intercept of a line. The LINREG program was initially introduced in Chap. 3 and is covered in detail in App. C.

The output of LINREG displays the slope and intercept of the low-I_{ds} I-V characteristics. The output of the LINREG program is shown in Fig. 5.22.

From the slope (b) and y-axis intercept (a) shown in Fig. 5.22, the V_{gs} intercept can be determined from Eq. 5.16.

$$V_{gs}(\text{intercept}) = x\text{-axis(intercept)} = -\frac{\text{a}}{\text{b}} = -3.27 \qquad (5.16)$$

Combining the result of Eq. 5.16 with Eq. 5.13 gives the value of VTO as shown in Eq. 5.17.

$$\text{VTO} = V_{gs}(\text{intercept}) - \frac{V_{ds}}{2} = -\,3.27 - \frac{0.1}{2} = -\,3.32\text{ V} \quad (5.17)$$

To calculate BETA, use the slope (b) from Fig. 5.22 and Eq. 5.14. The value of BETA will be set as shown in Eq. 5.18.

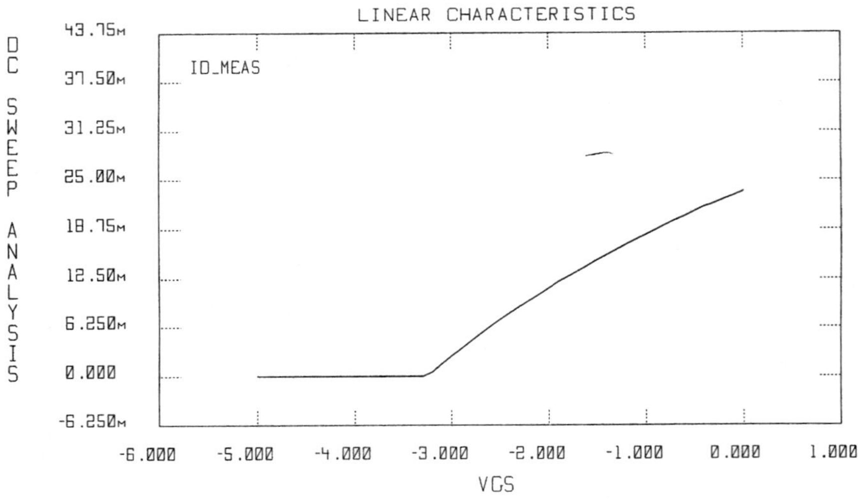

Figure 5.19 Measured linear characteristics.

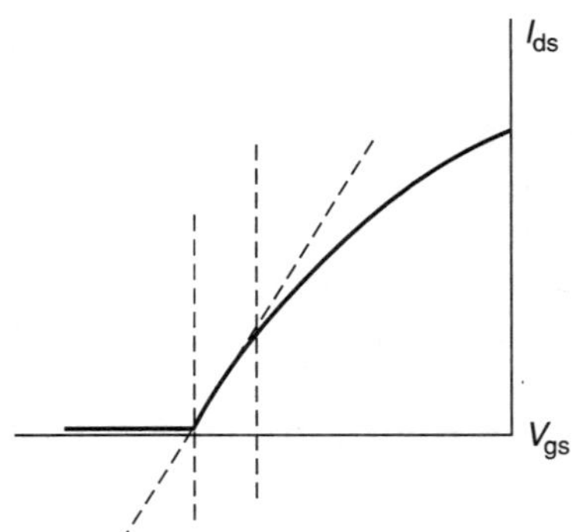

Figure 5.20 Extracting slope and intercept of linear characteristics.

Vgs	Ids
-3.000E+00	2.566E-03
-2.900E+00	3.533E-03
-2.800E+00	4.486E-03
-2.700E+00	5.413E-03

Figure 5.21 Input to LINREG program.

```
* * * * * * * * * * LINREG Results * * * * * * * *
*
* For the equation Y = a + b*X
*
* a    =    0.31057E-01
* b    =    0.94940E-02
* chi2 = 0.000
*
```

Figure 5.22 LINREG output.

$$\text{BETA} = \frac{\text{slope}}{(2^*V_{\text{ds}})} = \frac{0.9494\text{e-}2}{(0.2)} = 47.47\text{m mhos} \qquad (5.18)$$

To compare the measured data to the simulated model, we entered the newly calculated values of VTO and BETA into the NCHAN.MOD file in the RCV34\CH5 subdirectory. With the new model parameters entered into the NCNAN.MOD file, the linear characteristics of the model were simulated and merged with the measured data obtained from the HP 4142a parametric analyzer. The resulting curve is illustrated in Fig. 5.23.

Figure 5.23 demonstrates a good fit between the measured and simulated characteristics at low I_{ds} currents. As the current starts to increase, the series resistance of the drain and source begin to bend the measured characteristics away from the modeled (ideal) characteristics.

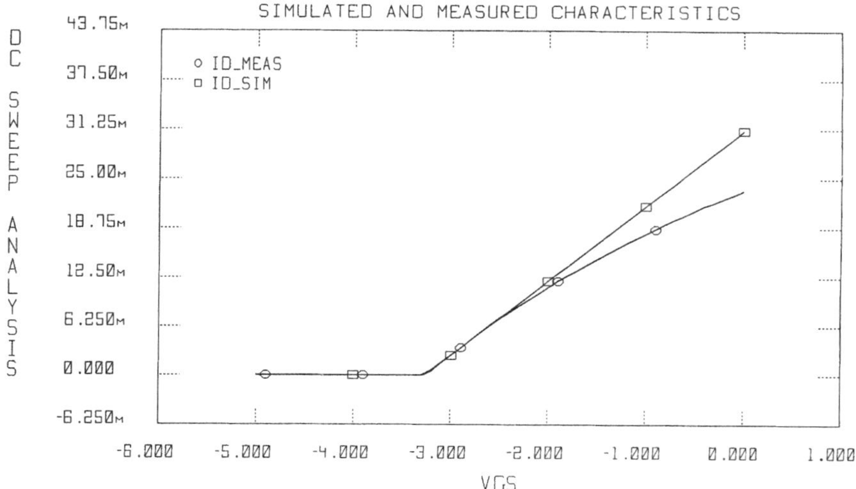

Figure 5.23 Measured and simulated characteristics after VTO and BETA extraction.

After obtaining accurate values of VTO and BETA, we are ready to use the RS_CALC program to determine the drain and source resistance. RS_CALC requires the user to enter the terminal voltages, V_{ds} and V_{gs}, the calculated VTO and BETA, and the measured drain current at $V_{gs} = 0$ and the theoretical drain current at $V_{gs} = 0$. All of these quantities are already known, except for the theoretical drain current at $V_{gs} = 0$.

To determine the theoretical drain current at $V_{gs} = 0$, we can either use the graphical technique illustrated in Fig. 5.15, or we can simply return to the simulation of the linear characteristics and read the theoretical current from the curve at $V_{gs} = 0$, as shown in Fig. 5.24(*a*). From this plot, the maximum theoretical current is 31.05 mA. This will be used as the value of $I_{d(high)}$ in the RS_CALC program. Figure 5.24(*b*) illustrates the value of current we will use for $I_{ds(low)}$ in RS_CALC.

Entering the values shown in Table 5.2 into the RS_CALC program produces the result shown in Fig. 5.25.

Simulating the JFET model after extracting VTO, BETA, RD, and RS and comparing the simulated and measured characteristics resulted in the plot of Fig. 5.26(*a*), in which the simulated and measured characteristics appear to fit well at very low drain currents and at very high drain currents, but between these limits, the accuracy is rather poor. Figure 5.26(*b*) exaggerates the problem and clearly illustrates that our model appears to have too little gain (BETA) when compared with the simulated characteristics. This is not an uncom-

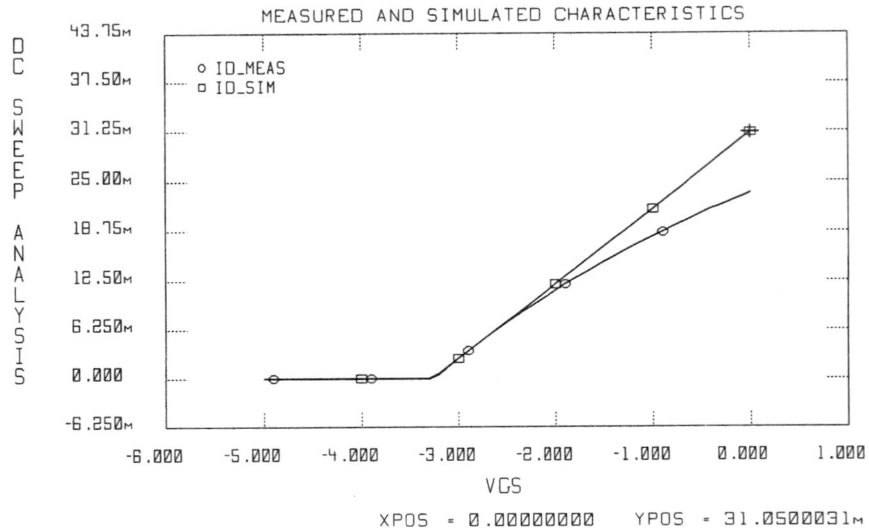

Figure 5.24(*a*) Determining the theoretical maximum drain current.

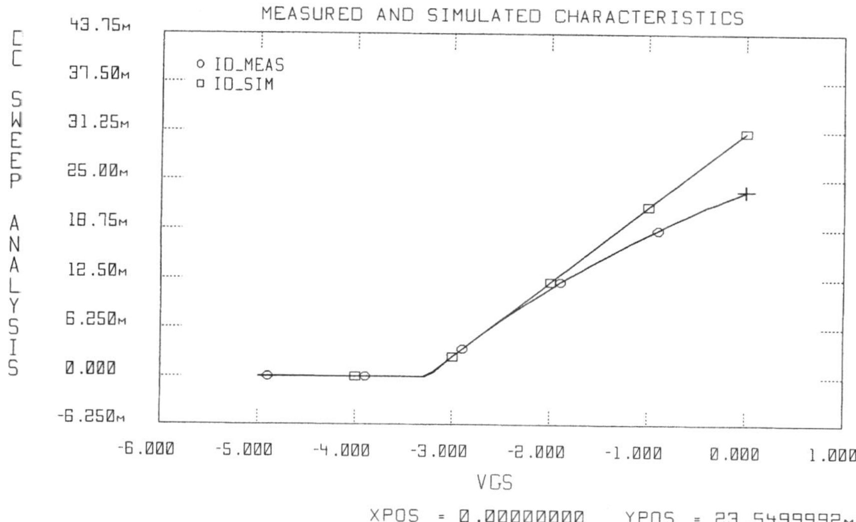

Figure 5.24(*b*) Determining the measured drain current at $V_{gs} = 0$.

TABLE 5.2 RS_CALC Program Inputs

Prompt	Value
V_{ds}	0.1
V_{gs}	0
VTO	−3.32
BETA	47.47e-3
$I_{ds(high)}$	31.05e-3
$I_{ds(low)}$	23.55e-3

```
* * * * * * * * * * RS_CALC Results * * * * * * * * *
*
* RD = RS =     4.99436E-01
*
*
* * * * * * * * * * * * * * * * * * * * * * * * * * *
```

Figure 5.25 RS_CALC output.

mon problem, and is due to the data points used to calculate VTO and BETA.

Again, rather than rerunning the LINREG program on a new set of data, we decided to adopt the practical approach to improving our model. To try to improve our model, the value of BETA was increased, and the model was resimulated. Several different values of BETA were used before finally selecting a value of 0.05 for the final model.

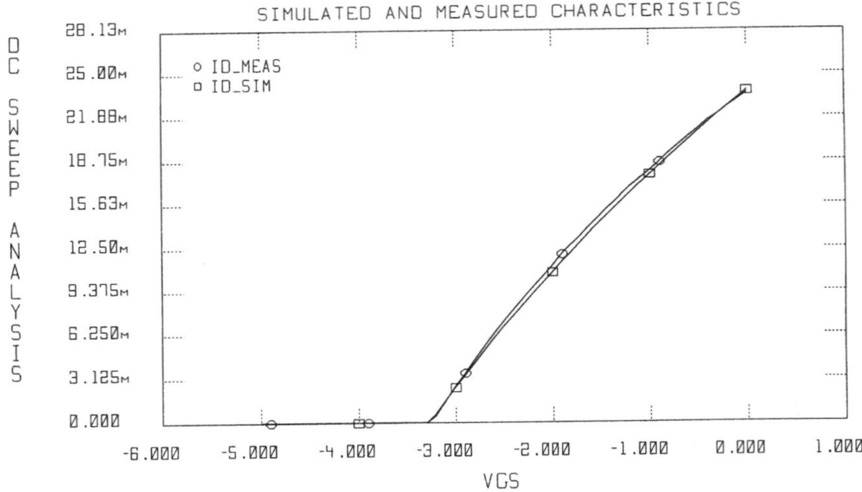

Figure 5.26(*a*) Measured and simulated characteristics after BETA, VTO, RS, and RD extraction.

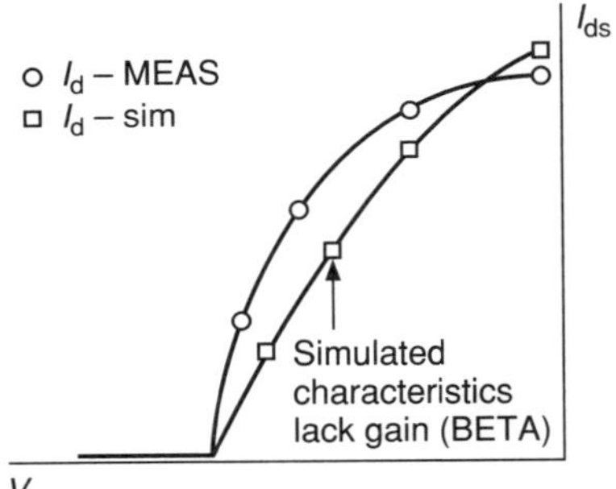

Figure 5.26(*b*) Measured and simulated characteristics (exaggerated view).

The simulation result with the "optimized" value of BETA is illustrated in Fig. 5.27.

While Fig. 5.27 demonstrates that our model fits well with measured data at both low and midrange drain currents, the high drain current curves don't compare well. This is due to the series resistance values calculated earlier. The values of RD and RS were selected for the old value of BETA. Since the value of BETA has changed, we should refit the parameters RD and RS. To do this, we need the new maximum value of current at $V_{gs} = 0$. This is easily done by setting the value of RD and RS to zero and simulating the linear characteristics of our JFET model. With the new value of BETA, the model produces a higher maximum current $I_{ds(high)}$, as shown in Fig. 5.28. From Fig. 5.28, the maximum current (at the marker) is 32.7 mA. This value of current, along with the other values of Table 5.2 were

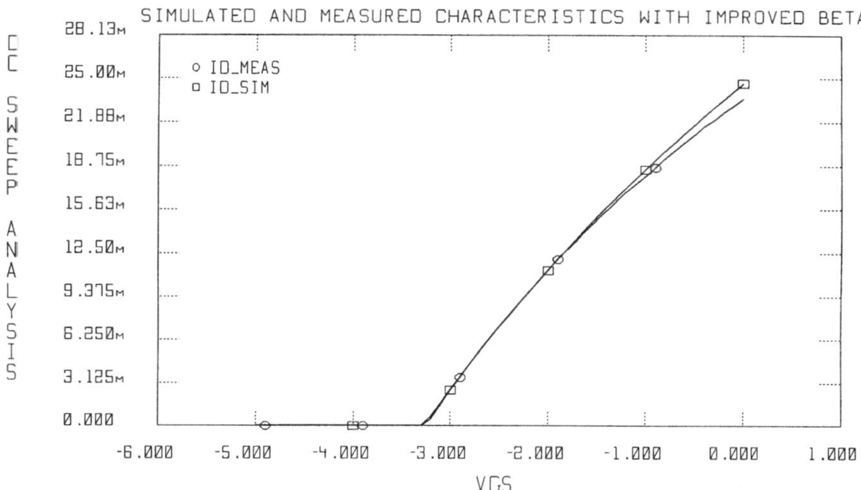

Figure 5.27 Measured and simulated characteristics with BETA = 0.05.

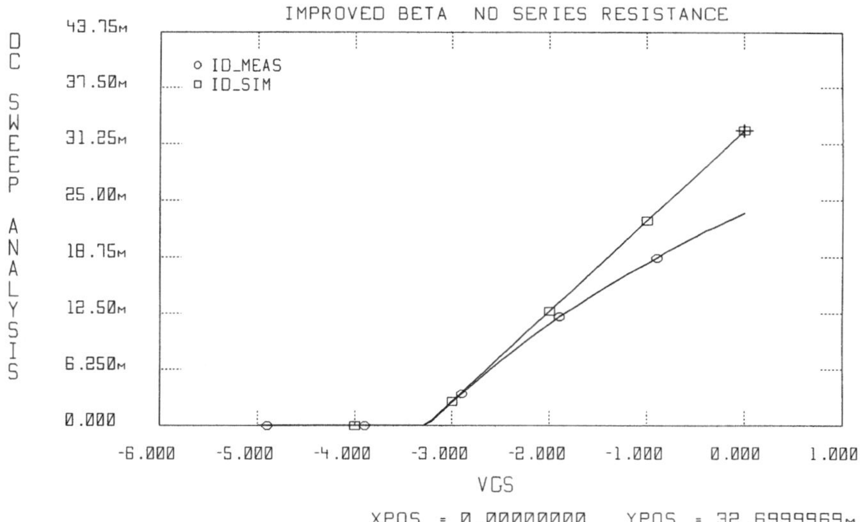

Figure 5.28 Determining a new value of $I_{ds(high)}$ by simulating with RS and RD set to zero.

entered into the RS_CALC program. The output of RS_CALC is shown in Fig. 5.29. The new values for RD and RS are 5.75 Ω.

When the new values of RS and RD are added to the model (along with the new value of BETA), the accuracy of the model improves dramatically. The improved model is illustrated in Fig. 5.30 and exhibits an rms error of less than 1.5 percent.

```
* * * * * * * * * * RS_CALC Results * * * * * * * * *
*
* RD = RS =     5.75529E-01
*
*
* * * * * * * * * * * * * * * * * * * * * * * * * * *
```

Figure 5.29 Newly calculated value of RD and RS.

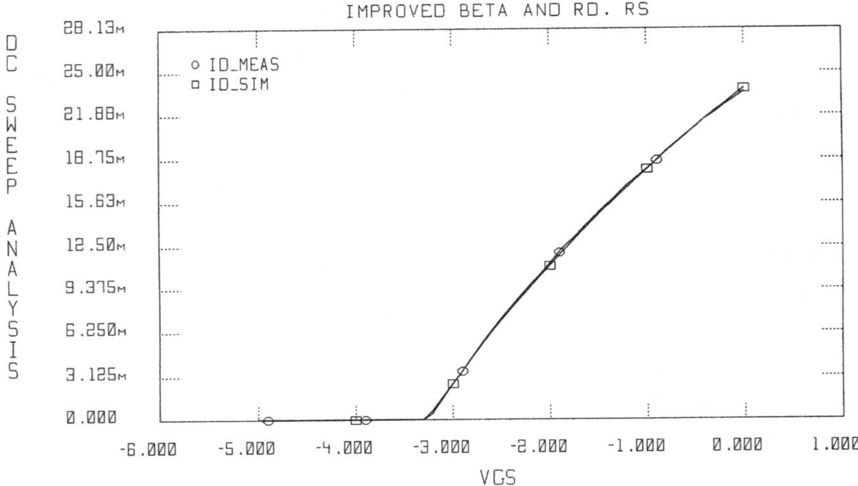

Figure 5.30 Simulated and measured characteristics with optimized values of BETA, RD, and RS.

The last significant piece of the DC model is the extraction of the parameter LAMBDA. LAMBDA is an empirical parameter which determines the slope of the output characteristics. Figure 5.31 illustrates the simulated and measured family of curves. While some may settle for this level of accuracy (about 6.7 percent rms error), others may not. To improve the model, notice how the measured output characteristics slope up just slightly, while our model remains flat in the saturated region. By adding LAMBDA, we can account for the slope of the output characteristics.

LAMBDA is determined from slope of the output characteristics. While the slope (and intercept) of the output curves could be determined graphically, we elected to use the LINREG program to improve the accuracy of our extraction. The measured I-V data between $V_{ds} = 4$ and $V_{ds} = 5$, illustrated in Fig. 5.32, was saved in a file and entered into the LINREG program. The actual data is shown in Fig. 5.33(*a*), and the result of the calculation is shown in Fig. 5.33(*b*). Using these results and Eq. 5.9 yields the value of LAMBDA.

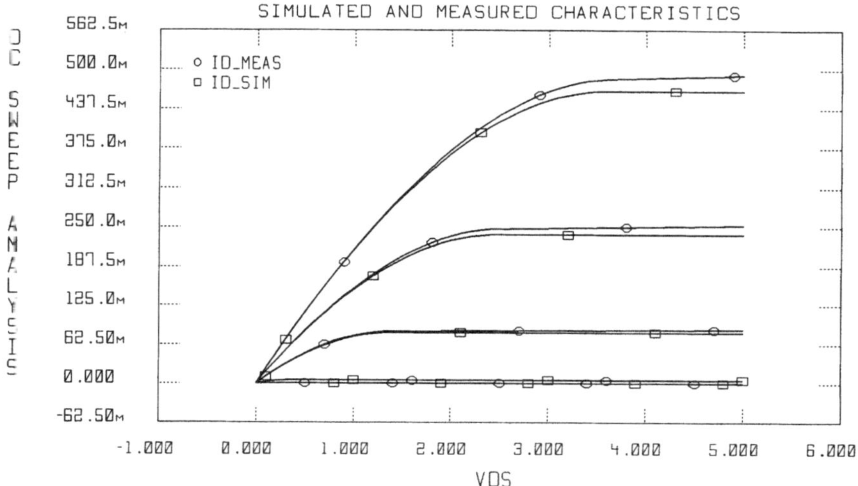

Figure 5.31 Measured and simulated characteristics (before LAMBDA extraction).

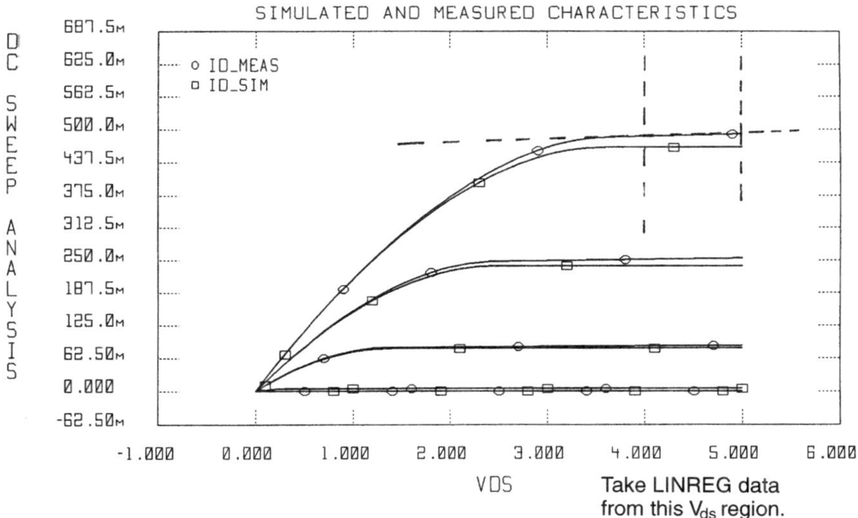

Figure 5.32 Data for LINREG program falls between V_{ds} = 4 and 5 V.

$$\text{LAMBDA} = \frac{\text{slope}}{y\text{-intercept}} = \frac{0.4354\text{E-2}}{0.46925} = 0.00928 \tag{5.19}$$

The final DC parameter is the gate (diode) leakage parameter, IS. While IS could be measured (the parameter has very little influence on the normal I-V curves), we will use the data sheet value for IS: I_{GSS}. On the data sheet, I_{GSS} has a typical value of 5 pA. This is the value we will use for IS.

```
Vds            Ids

4.000E+00      4.867E-01
4.100E+00      4.871E-01
4.200E+00      4.875E-01
4.300E+00      4.880E-01
4.400E+00      4.884E-01
4.500E+00      4.888E-01
4.600E+00      4.893E-01
4.700E+00      4.897E-01
4.800E+00      4.902E-01
4.900E+00      4.906E-01
5.000E+00      4.910E-01
```

Figure 5.33(*a*) Measured Data for LAMDA extraction.

```
* * * * * * * * * * LINREG Results * * * * * * * *
*
* For the equation Y = a + b*X
*
* a    =    0.46925E+00
* b    =    0.43546E-02
* chi2 = 0.000
*
```

Figure 5.33(*b*) LINREG results.

Figure 5.34 illustrates our final DC model. The calculated rms error between the simulated family of curves and the measured family of curves is a very acceptable 3.03 percent.

With the DC parameters determined, we can focus our attention on the capacitive parameters. The two most important capacitive parameters are CGD and CGS. These represent the zero-biased gate-drain and gate-source capacitance of the JFET. While the data sheet gives values for these, the values stated in the data sheet are determined with a bias applied between the drain and source. The model parameters CGD and CGS must be determined with no bias applied. For this reason, the data sheet information will not serve our needs, and we must again turn to our capacitance meter to measure these values.

Since most JFETs are symmetric, and because CGS and CGD are in parallel with each other, we can simplify our measurement procedure by measuring both CGS and CGD at the same time. Figure 5.35(*a*) illustrates the measurement setup to determine CGS and CGD.

Because the capacitances CGS and CGD are actually junction capacitances of the gate diffusion, changing the bias on the capacitance tester produces a typical junction (diode) capacitance curve. Figure 5.35(*b*) illustrates a typical junction capacitance curve

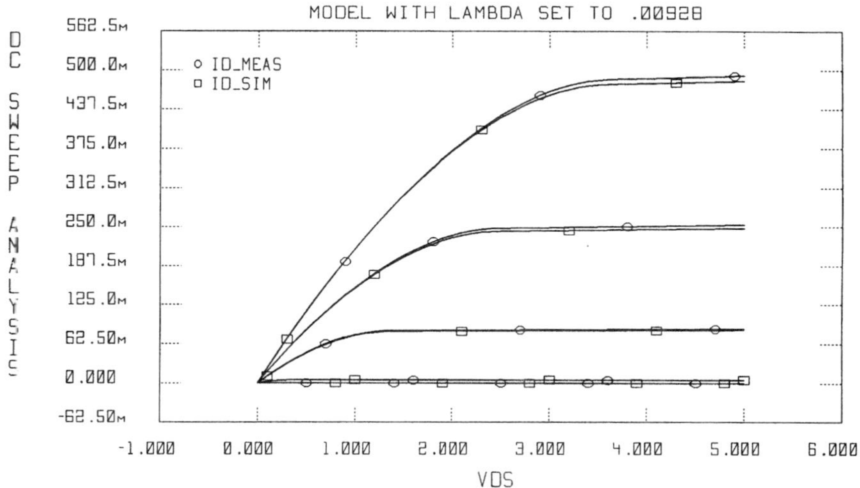

Figure 5.34 Measured and simulated data with LAMBDA set to 0.00928.

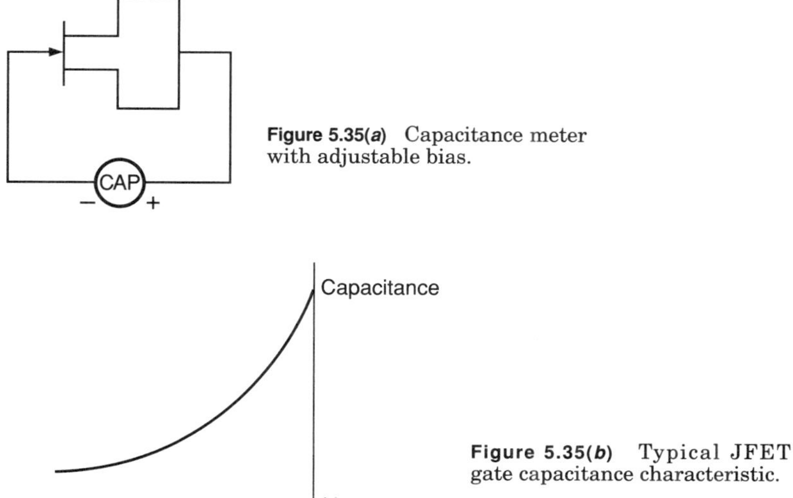

Figure 5.35(*a*) Capacitance meter with adjustable bias.

Figure 5.35(*b*) Typical JFET gate capacitance characteristic.

obtained from measuring CGS and CGD. The model parameters CGS, CGD, and PB (VJ) can be determined with the help of CJ_CALC.

CJ_CALC is a program which calculates the junction capacitance parameters for any junction capacitance voltage characteristic. Using CJ_CALC is discussed in detail in App. B.

To use CJ_CALC to determine the capacitance model parameters, we must measure the junction capacitance at several different reverse-bias voltages. (Suggested voltage settings are presented in App. B.) After measuring the capacitance values, the measured data is entered into the CJ_CALC program. CJ_CALC determines the values of the parameters CJO, M, and VJ, and these correspond to the values required for our model, as shown in Fig. 5.36.

To use the parameters calculated by CJ_CALC, set PB to the value of VJ, and set CGD to half the value of CJO, and set CGS to half the value of CJO. CGD and CGS must be set to half of the measured value because when they were measured, you actually measured both capacitances.

Figure 5.37(*a*) illustrates the measured and simulated gate capacitance of the JFET, Fig. 5.37(*b*) displays the model parameters calculated by CJ_CALC, and Fig. 5.37(*c*) shows the final JFET model capacitance parameter values. The error in Fig. 5.37(*a*) can be attributed to the inability to set the model parameter M for the JFET. Many of the vendor-offered SPICE-like simulators give the user the ability to adjust M, but standard SPICE does not.

Parameters from CJ_CALC	JFET parameters
CJO	CGS and CGD
M	not used
VJ	PB

Figure 5.36 JFET capacitance parameters.

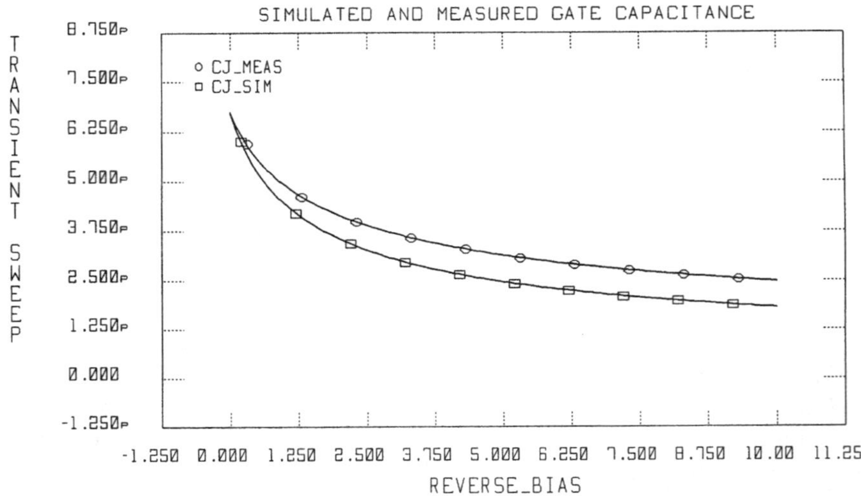

Figure 5.37(*a*) Measured and simulated gate capacitance.

CJO = 6.74 pf
M = 0.38
VJ = 0.76

Figure 5.37(*b*) CJ_CALC output.

CGS = 3.37 pf
CGD = 3.37 pf
PB = 0.76

Figure 5.37(*c*) JFET capacitance parameters.

VTO	=	−3.32 V
BETA	=	0.05 A/V^2
LAMBDA	=	0.00928 V^{-1}
RD	=	0.575 Ω
RS	=	0.575 Ω
IS	=	5 pA
CGS	=	3.37 pf
CGD	=	3.37 pf
PB	=	0.76
FC	=	0.5

Figure 5.38 Final JFET model parameter values.

Figure 5.38 illustrates the complete JFET model parameter set. Like the other device models in SPICE, the JFET parameters are easy to obtain once you understand how they are measured. While the DC parameters fit well, the capacitance parameters, due to the inability to set M, give only a fair model of capacitance.

References

1. Walter Curtice, "A MESFET Mode for Use in the Design of GaAs Integrated Circuits," *IEEE Trans. Microwave Theory and Techniques,* vol. MTT-28, May 1980, pp. 448–456.
2. H. Shichman, and D. Hodges, "Modeling and Simulation of Insulated-Gate Field Effect Transistor Switching Circuits," *IEEE Journal of Solid-State Circuits,* SC-3, 1968, pp. 85–89.

Chapter

6

Modeling the Power MOSFET

Introduction

In this chapter, you will learn to model the power metal-oxide-silicon field-effect transistor (power MOSFET). In modeling the power MOSFET, we will adopt a slightly different approach than we have in the previous three chapters, because modeling power MOSFETs accurately cannot be accomplished with the simple SPICE MOSFET primitive. Power MOSFETs are vertical structures, where the SPICE MOSFET primitive models the characteristics of a lateral transistor. Figure 6.1(*a*) and (*b*) illustrates the differences between the vertical power MOSFET and the lateral small-signal device described by SPICE.

To accurately model the power MOSFET requires the use of a subcircuit model or macromodel. Since 1982,[1–6] different macromodels

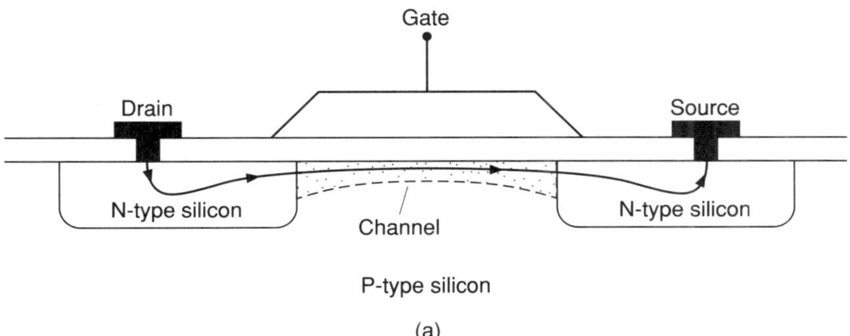

Figure 6.6(*a*) Small-signal (lateral) N-channel MOSFET cross section showing direction of current flow.

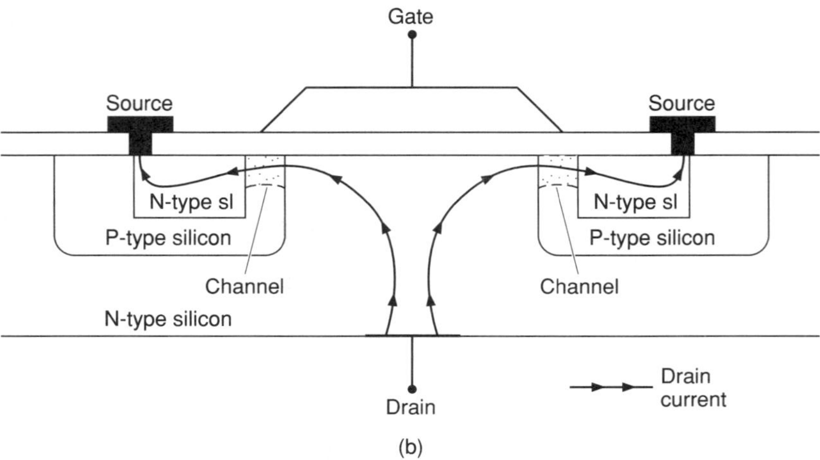

Figure 6.1(*b*) Power (vertical) N-channel MOSFET cross section showing direction of current flow.

have been proposed to improve the accuracy of the standard small-signal MOSFET in SPICE. Unfortunately, most of these models were either too simplistic to model all of the desirable characteristics of the power MOSFET or too complex (and often suffering from nonconvergence problems) for large simulations.

One power MOSFET macromodel which exhibits accurate I-V, C-V, and gate-charge characteristics while remaining a relatively simple macromodel was created by Richard Williams and Iraj Masarrati of Siliconix.[7] Although quite accurate, this model is simple in structure and does not rely on capacitor-bank switching which may lead to nonconvergence problems.

Because the Siliconix macromodel displays superior accuracy to the previous macromodels, exhibits good convergence characteristics, and is simple in structure, the Siliconix macromodel is the macromodel we will use to model power MOSFETs.

The Siliconix macromodel is shown in Fig. 6.2. At the core of the macromodel is a standard SPICE MOSFET transistor. Around this core, other SPICE elements are added. The devices which are added to the MOSFET core of the model include D_{BODY}, the body diode; CGS, the gate-source capacitance; CGD, the fixed gate-drain capacitance; LD, LG, and LS, the series inductance of the device terminals; R_D, the temperature-dependent drain resistor; and D_{MIN} and D_{MAX}, diodes to help model the nonlinear gate-drain capacitance and gate-charge characteristics.

We will begin modeling this device by describing the parameters of the MOSFET transistor, then the external diodes. Each of these parameters will be described in detail. We will continue by describing

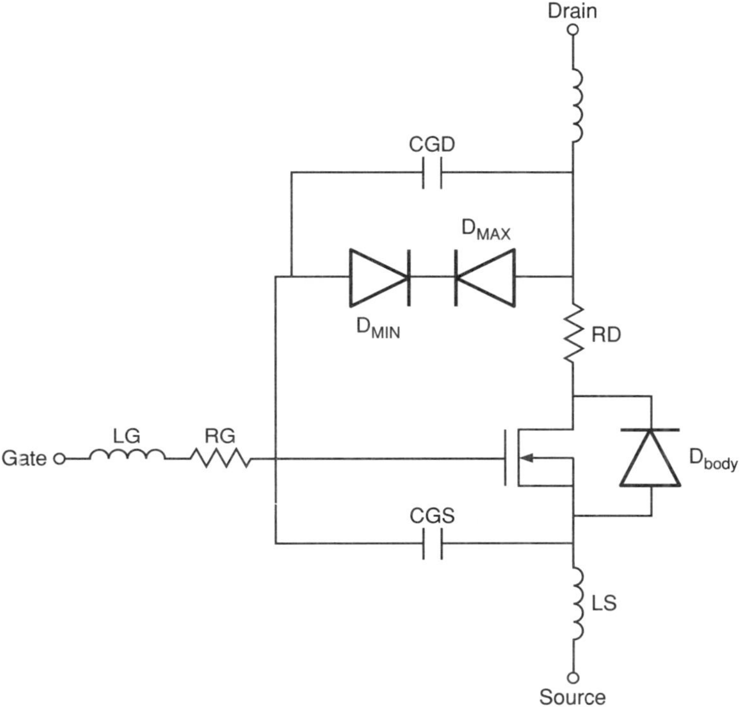

Figure 6.2 Siliconix N-channel power MOSFET model.

functionality of the macromodel and how each of the devices work together. The behavior of the macromodel will be compared with the information typically found on power MOSFET data sheets. Once the model has been explained, extraction techniques will be presented to calculate the parameter values of each element in the model. To further illustrate the extraction procedures, a Motorola MTD5N06 Power MOSFET will be modeled.

Just as a reminder, if you have already read Chap. 3 of this text, regarding modeling junction diodes, you are now ready to model a power MOSFET. If you skipped Chap. 3 and turned directly to this section, you should first read the section on modeling the diode. The information learned about the diode parameters and the modeling and extraction techniques covered in the diode chapter will be used in this chapter and others. The diode chapter also introduced the programs CJ_CALC and LINREG. We will make extensive use of these programs throughout this chapter too.

Throughout this chapter, various circuits and models will be described. All of the circuit and models are found in the CH6 subdirectory of the disk bound to this text. The utility programs CJ_CALC,

CJB, LINREG, and LRB, and RS_CALC are found in the UTIL subdirectory of the disk bound to this book. The disk must be installed on your hard disk and uncompressed before you can access these files.

Modeling Approach

In keeping with our philosophy of practical device modeling, a typical power MOSFET data sheet will be inspected for information which might be useful in developing the device model. What cannot be obtained from the data sheet will be taken from measured I-V and C-V characteristics. After the initial set of model parameters has been chosen, simulations will be run on the model to optimize the final parameter values and to estimate the accuracy of our model.

Power MOSFET Model Parameters

At the core of the power MOSFET macromodel is a standard SPICE small-signal MOSFET. The DC I-V parameters of this transistor can be adjusted to accurately model the I-V behavior of the power MOSFET. The DC parameters of the MOSFET are listed in Table 6.1.

Table 6.1 illustrates only 5 of the 42 MOSFET model parameters available in SPICE.[8] The reason we have dismissed 37 transistor parameters is that many of the model parameters of the standard SPICE MOSFET describe the short-channel behavior of the transistor or capacitive behavior of the transistor. Short-channel effects rarely influence the behavior of power MOSFETs and, therefore, many of these parameters have been discarded. Furthermore, since the MOSFET at the core of the macromodel is used to describe the DC I-V behavior only, the capacitive model parameters of the standard MOSFET have also been eliminated. Rather than use the capacitive model parameter of the MOSFET, the capacitive characteristics of the power MOSFET macromodel will be described by the external devices of the macromodel.

Table 6.2 illustrates the model parameters of the external devices which must be determined for the power MOSFET macromodel, while

TABLE 6.1 MOSFET Model Parameters

Name	Parameter	Units	Default
LEVEL	Model index	—	1
VTO	Zero-bias threshold voltage	V	0.0
KP	Transconductance parameter	A/V^2	2.0E-5
LAMBDA	Channel-length modulation	1/V	0.0
RS	Source ohmic resistance	Ω	0.0

TABLE 6.2 Power MOSFET Macromodel Elements

(Active elements external to the core MOSFET)			
Name	D_{BODY}	D_{MIN}	D_{MAX}
Parameters	IS		
	RS		
	N		
	CJO	CJO	CJO
	VJ	VJ	
	M	M	

Table 6.3 illustrates the passive elements which are used in the power MOSFET macromodel.

The model parameters of Tables 6.1 and 6.2, and the element values of Table 6.3, describe the I-V and C-V characteristics of the power MOSFET macromodel. The first five parameters (the DC parameters of the standard SPICE MOSFET) include LEVEL, VTO, KP, LAMBDA, and RS. These parameters describe the DC characteristics of the power MOSFET. (For many power MOSFETs, the Level 1 transistor equations can be used. To improve the accuracy of some low-voltage devices, readers may want to switch to the Level 3 model and add the model parameters THETA and VMAX.[8]

The diode parameters of D_{BODY} simulate the characteristics of the body diode. While the SPICE MOSFET has a built-in body diode, the built-in model has very little flexibility and cannot be easily matched with measured characteristics. The D_{BODY} diode overcomes these problems.

The diodes D_{MIN} and D_{MAX} help model the nonlinear gate-drain capacitance and the gate-charge characteristics.

The LS, LD, and LG inductors model the lead inductance of the transistor. RG represents the gate or spreading resistance of the device. RD represents the drain resistance of the MOSFET. In this model, an external RD is used to model the drain resistance because the built-in drain resistance of the MOSFET does not allow the user to describe the temperature coefficients of the transistor. The external resistance RD can be adjusted to change with temperature. The capacitors CGD and CGS represent the constant gate-drain and gate-source capacitances.

TABLE 6.3 Power MOSFET Macromodel Elements

(Passive elements external to the core MOSFET)						
Name						
LS	LD	LG	RG	RD	CGD	CGS

The DC power MOSFET parameters

The most basic and most important parameters of the power MOSFET model are VTO, KP, and LAMBDA. VTO represents the gate threshold voltage of the transistor. VTO determines the gate voltage at which the transistor begins conducting current. KP is the transconductance of the device. To better understand these parameters, the equations used to describe the drain current of the MOSFET are shown in Eqs. 6.1, 6.2, and 6.3. Figure 6.3 illustrates the two regions of operation described by Eqs. 6.1 and 6.2.

$$I_{ds} = \mathrm{KP} * \left(V_{gs} - \mathrm{VTO} - \frac{V_{ds}}{2}\right) * V_{ds} * (1 + \mathrm{LAMBDA} * V_{ds}) \quad (6.1)$$

for linear region $|V_{gs} - \mathrm{VTO}| > V_{ds}$
and $0 < |V_{gs}| < |\mathrm{VTO}|$

$$I_{ds} = \frac{\mathrm{KP}}{2} * (V_{gs} - \mathrm{VTO})^2 * (1 + \mathrm{LAMBDA} * V_{ds}) \quad (6.2)$$

for saturated region $V_{ds} >= |V_{gs} - \mathrm{VTO}|$
and $0 < |V_{gs}| < |\mathrm{VTO}|$

$$I_{ds} = 0 \quad (6.3)$$

for cutoff region $|V_{gs}| < |\mathrm{VTO}|$

From Eqs. 6.1 and 6.2, KP represents the proportionality constant between the drain current and the terminal voltages. VTO represents the voltage at which the transistor begins conducting current. Both of

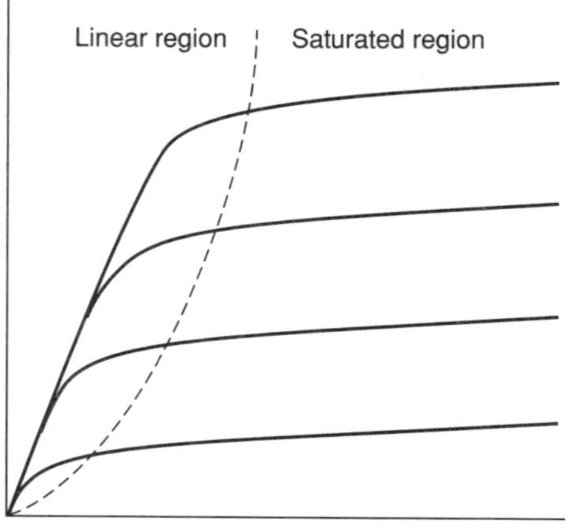

Figure 6.3 Linear and saturated regions of operation for a MOSFET.

these parameters are determined from the I-V characteristics of the transistor.

The MOS I-V characteristic familiar to most designers is the family of curves. A typical family of curves is shown in Fig. 6.3. Unfortunately, the family of curves does not provide us with enough information to determine KP and VTO. To extract these two parameters, we need a different set of I-V characteristics. Figure 6.4(*a*) illustrates another set of I-V characteristics from the JFET. These characteristics are known as the *linear* characteristics of a MOSFET. The linear characteristics get their name from the fact that, to obtain this curve, the transistor must remain biased in the linear region of operation. Figure 6.4(*b*) illustrates the measurement setup for the linear characteristics. From the linear characteristics, we will obtain model parameters VTO, KP, RS, and a value for the RD resistor.

Figure 6.5(*a*) illustrates the saturated characteristics of the MOSFET. Although similar in appearance to the linear characteristics, the saturated characteristics are obtained with the transistor biased in the saturated region of operation. [While Figs. 6.4(*a*) and 6.5(*a*) are similar in appearance, Fig. 6.5(*a*) has a parabolic shape where Fig. 6.4(*a*) has a more linear shape. The difference in the graphs can be seen if you hold the book at an acute angle, with the bottom of the book closest to you and the top of the book farthest from you. In this position, you can see the curvature of Fig. 6.5(*a*).] The difference between the linear and saturated characteristics is found in the measurement setup shown in Fig. 6.5(*b*). The measurement setup in Fig. 6.4(*b*) holds V_{ds} to a small voltage value (100 mV). Holding V_{ds} below V_{gs} forces the transistor to operate in the linear region only. The measurement setup in Fig. 6.5(*b*) holds V_{ds} and V_{gs} at the same potential. By keeping V_{ds} greater or equal to V_{gs}, the transistor is forced to oper-

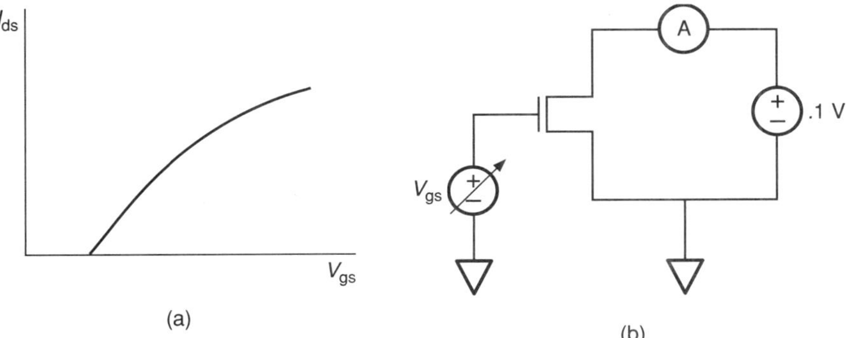

Figure 6.4 Linear characteristics of a MOSFET and measurement setup for linear characteristics.

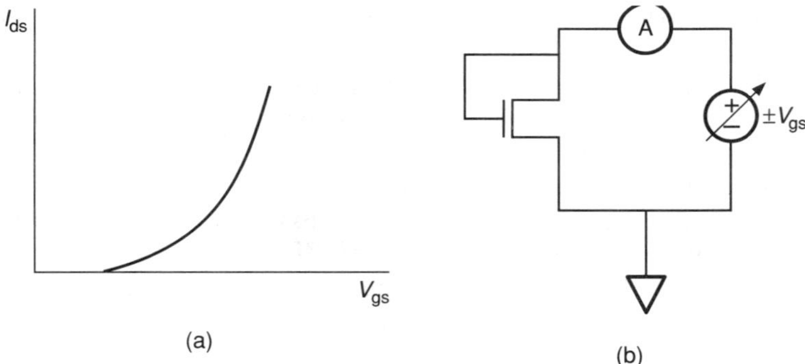

Figure 6.5 Saturated characteristics of a MOSFET and measurement setup for saturated characteristics.

ate in the saturated region of operation. The drain current in Figs. 6.4(*a*) and 6.5(*a*) are the drain currents described by Eqs. 6.1 and 6.2, respectively.

Of the three standard I-V relationships of the MOSFET (family of curves, linear characteristics, saturated characteristics), the standard power MOSFET data sheet will usually contain the saturated characteristics, sometimes contain the family of curves, and almost never contain the linear characteristics. This is unfortunate because most of our modeling information comes from the linear characteristics.

However, before we go further, three programs have been created which help you simulate the linear, saturated, and family of curves of a MOSFET. The first of these programs is called LIN.BAT. LIN.BAT is a batch procedure which runs RSPICE on a file named LIN.CIR. The LIN.CIR file is a SPICE input file which simulates the linear characteristics of a MOSFET. The LIN.CIR file reads the MOSFET model parameters from a file named PMOS.MOD (the P stands for power not P-channel). The output from the LIN.CIR simulation is shown in Fig. 6.6(*a*). The program SAT.BAT is similar to LIN.BAT but generates the saturated characteristics of the MOSFET. The output of SAT.BAT is shown in Fig. 6.6(*b*). The final program, FAMILY.BAT simulates the family of curves of a MOSFET. The output of FAMILY.BAT is shown in Fig. 6.6(*c*). All three of these programs are stored in the `RCGV34\CH6` subdirectory. To run these programs, change into the CH6 subdirectory and, at the DOS prompt, type:

```
LIN
```

or

```
SAT
```

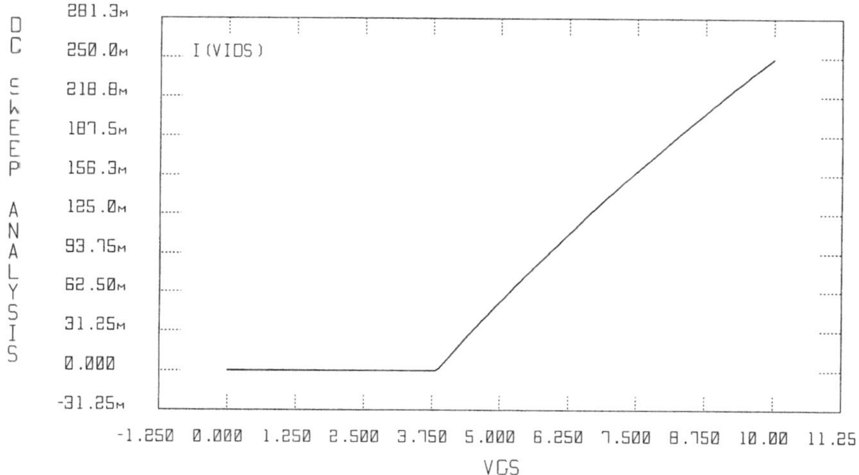

Figure 6.6(*a*) Output of LIN.BAT program.

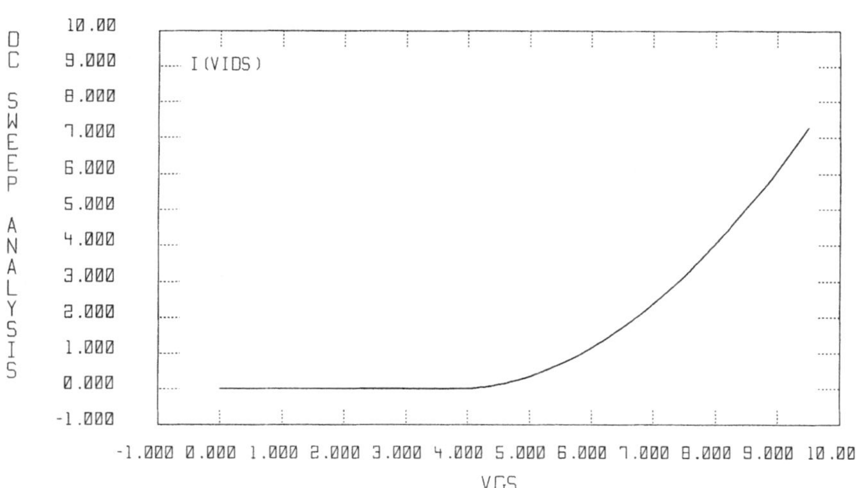

Figure 6.6(*b*) Output of SAT.BAT program.

or

```
FAMILY
```

The resulting RGRAPH plots should match Fig. 6.6(*a*), (*b*), and (*c*).

We will begin our discussion by illustrating techniques to determine VTO and BETA from the linear characteristics of a MOSFET. The reason the linear characteristics are preferred over the saturated

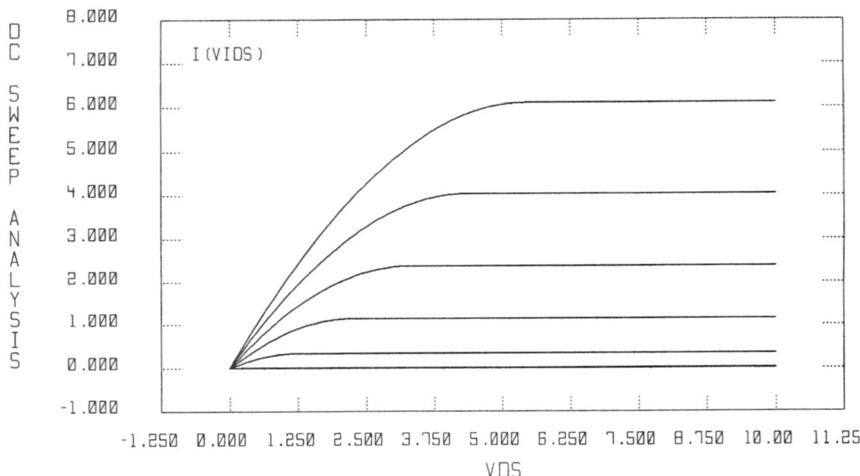

Figure 6.6(*c*) Output of FAMILY.BAT program.

characteristics is that the linear extraction method gives us a way to determine the source resistance RS and the drain resistance RD (and THETA if we elect to choose the Level 3 MOS equations). Neither of these resistance values can be determined from the saturated characteristics.

As with the JFET, the parameters VTO and KP will be obtained from the linear characteristics of the MOSFET. To determine VTO and KP make a straight-line approximation to the characteristics at low drain currents, as shown in Fig. 6.7. The intersection of this line and the V_{gs} axis is used to calculate VTO, the slope of the line is used to calculate BETA, and the extension of the line will be used to determine RD and RS. VTO is found from the intersection with the x-axis and Eq. 6.4. KP is calculated from the slope of the line and Eq. 6.5.

While the slope and intercept may be determined graphically, a more accurate result is obtained with the LINREG program. LINREG

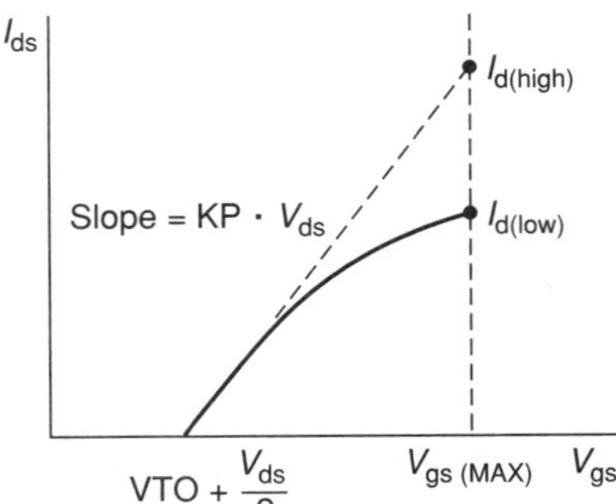

Figure 6.7 Extracting KP, VTO, RD, and RS from linear MOSFET characteristics.

is a linear regression program which calculates the slope and intercept of a line segment. By entering the linear I-V characteristics (after the transistor begins to turn on) into the LINREG program, the intercept and the slope of the curve are easily determined. (A complete user's guide to the LINREG program is found in App. C.)

$$\text{VTO} = V_{\text{gs(intercept)}} \frac{-V_{\text{ds}}}{2} \tag{6.4}$$

$$\text{KP} = \frac{\text{slope}}{V_{\text{ds}}} \tag{6.5}$$

To determine RS and RD, we must measure the drain current at $V_{gs} = V_{gs(MAX)}$. This current is known as $I_{d(low)}$. We also need to know the drain current at which the straight-line approximation would intersect the same V_{gs} voltage. This current is known as $I_{d(high)}$. Using these two currents, we can determine the drain and source resistance values.

$$I_{\text{d(high)}} - I_{\text{d(low)}} = 4 * \text{BETA} * I_{\text{d(low)}} * \text{RS} * \left(-V_{\text{ds}} - V_{\text{gs}} - I_{\text{d(low)}} * \text{RS} + \text{VTO} + \frac{V_{\text{ds}}}{2}\right) \tag{6.6}$$

Finding the source resistance involves solving Eq. 6.6, which is a quadratic, in terms of RS. To ease this calculation, a simple FORTRAN program has been created which solves Eq. 6.6 for RS. The program is called RS_CALC and is stored in the UTIL subdirectory of the disk which accompanies this text. The RS_CALC program asks the user for the terminal voltages of the device, the calculated VTO and BETA, and the $I_{d(high)}$ and $I_{d(low)}$ currents. The program solves Eq. 6.6 and determines the correct RS value.

Learning to use the RS_CALC program is beyond the scope of this chapter. The interested reader is directed to App. E for more information on using RS_CALC.

While some low-voltage power MOSFETs are fabricated symmetrically, most power MOSFETs are vertical structures and have an asymmetric drain and source resistance. There is, unfortunately, no simple test which will discriminate between the drain and source resistances. *For this reason, even though we may lose some accuracy, RD will be set equal to RS. Note:* For our power MOSFET model, the series resistance of the source is set with the RS model parameter, but the series resistance of the drain is set with the external resistor RD. This is done so the temperature coefficient of RD may be specified.

The body diode DC parameters

Accurately modeling the body diode parameters is an important aspect of the power MOSFET macromodel. To extract the body diode parameters, we will follow the same diode extraction procedures introduced in Chap. 3 of this text. The DC parameters we need to extract include IS, RS, and N for the body diode. For those not familiar with the extraction techniques of the SPICE diode, read Chap. 3. For those who have already read Chap. 3, Fig. 6.8(*a*) and (*b*) illustrates the measurement setup and extraction techniques for IS, RS, and N of the body diode.

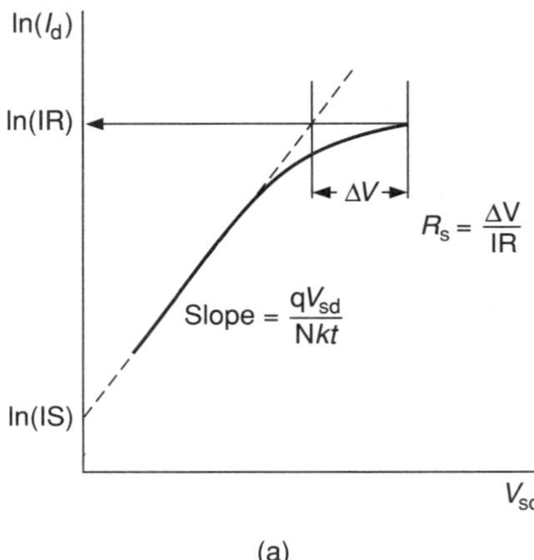

Figure 6.8(*a*) Extracting IS, RS, and N for the body diode.

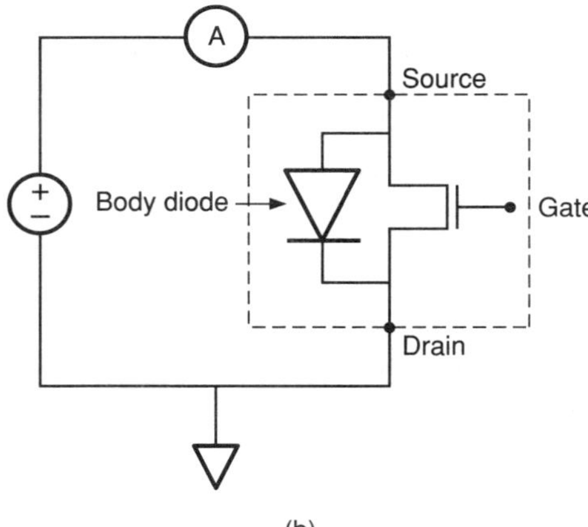

Figure 6.8(*b*) Measurement setup for body diode parameter extraction.

The capacitive power MOSFET parameters

Due to the vertical nature of the power MOSFET, the small-signal capacitance model of the MOSFET will not provide an accurate capacitance model for the power MOSFET. For this reason, the capacitance parameters of the MOSFET will be set to zero and the capacitance model will be constructed from other components.

To understand the capacitance model of the power MOSFET, we must return to the data sheet. The three traditional capacitances of interest are the input capacitance C_{iss}, the output capacitance C_{oss}, and the reverse-transfer capacitance C_{rss}. These three capacitances are illustrated in Fig. 6.9. Another plot which is important to the capacitance model is the gate-charge characteristics, shown in Fig. 6.10. The gate-charge characteristics illustrate the gate voltage vs.

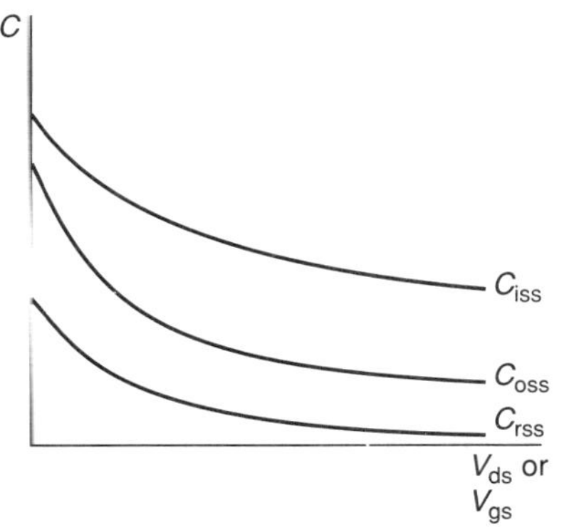

Figure 6.9 Capacitance characteristics of power MOSFET.

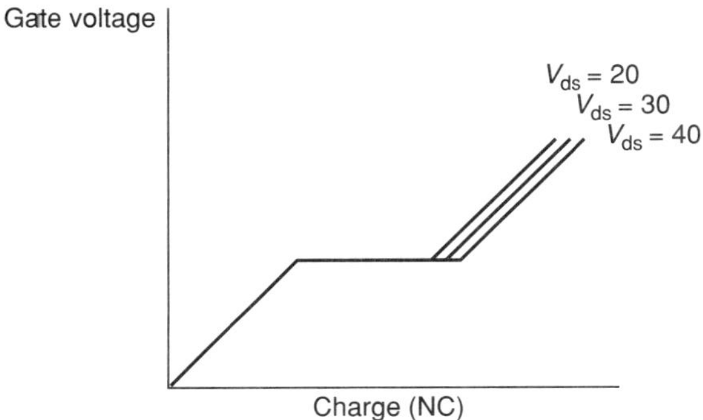

Figure 6.10 Gate charge characteristics of power MOSFET.

charge on the gate. Usually this plot is shown for three different V_{ds} voltages. Our power MOSFET model needs to accurately model both the capacitance characteristics and the gate-charge characteristics. *Both the capacitance characteristics and the gate-charge characteristics are a part of the information found on the power MOSFET data sheet.*

To help build our power MOSFET macromodel, two additional programs or circuit files have been developed. The first of these two is called GATE.BAT. The GATE.BAT batch file simulates the gate-charge characteristics of the power MOSFET. The GATE.BAT program reads the GATE.CIR circuit file which reads the parameters from the PMOS.MOD circuit file. The output of GATE.BAT is shown in Fig. 6.11(*a*) and the equivalent circuit used to generate the characteristics is shown in Fig. 6.11(*b*).

The second program is called CAP.BAT. CAP.BAT runs RSPICE on a file named CAP.CIR. The CAP.CIR circuit file reads from the PMOS.MOD circuit file. After RSPICE has completed the simulation of CAP.BAT, a program called PLOT_CAP.EXE reads through the CAP.OUT output file and reformats the data into a file which RGRAPH can read. The output of CAP.BAT is shown in Fig. 6.12.

Using the GATE and CAP programs will allow us to ensure the accuracy of our power MOSFET macromodel.

Functionally, the capacitance model of the power MOSFET macromodel can be shown as illustrated in Fig. 6.13. Unfortunately, we cannot measure the capacitances C_{gs}, C_{gd}, and C_{ds} directly; instead, these values must be obtained from C_{iss}, C_{oss}, and C_{rss}. The measurements

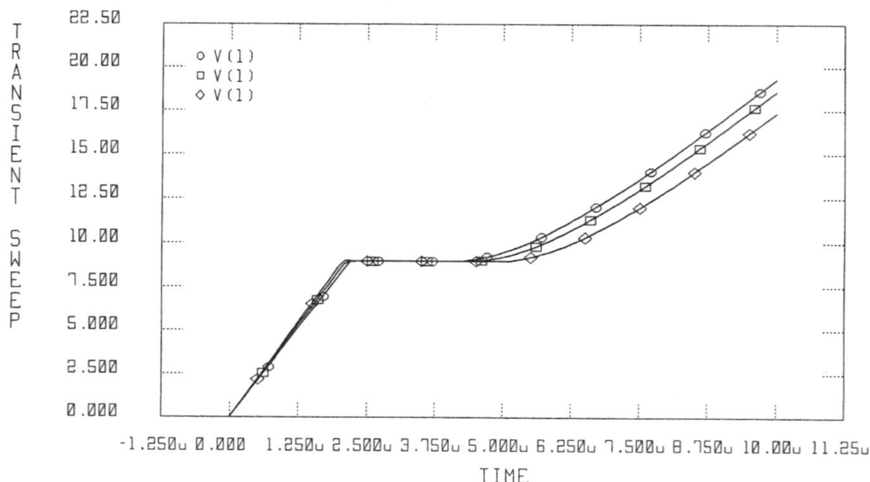

Figure 6.11(*a*) Output of GATE.BAT program. *Note:* x-axis scale is 1 μS = 1 nC.

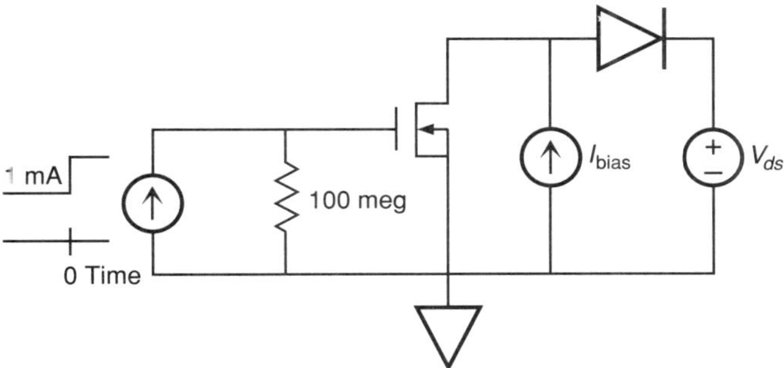

Figure 6.11(*b*) Gate-charge test circuit.

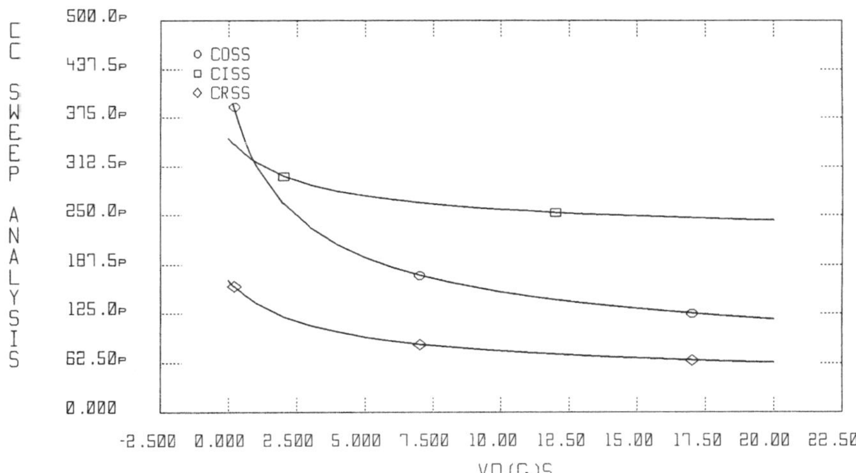

Figure 6.12 Output of CAP.BAT program.

for C_{iss}, C_{oss}, and C_{rss} are shown in Figs. 6.14(*a*), (*b*), and (*c*), and the relationships between C_{iss}, C_{oss}, C_{rss}, and C_{gs}, C_{gd}, and C_{ds} are shown in Eqs. 6.7, 6.8, and 6.9.

$$C_{iss} = C_{gs} + C_{gd} \tag{6.7}$$

$$C_{oss} = C_{gd} + C_{ds} \tag{6.8}$$

$$C_{rss} = C_{gd} \tag{6.9}$$

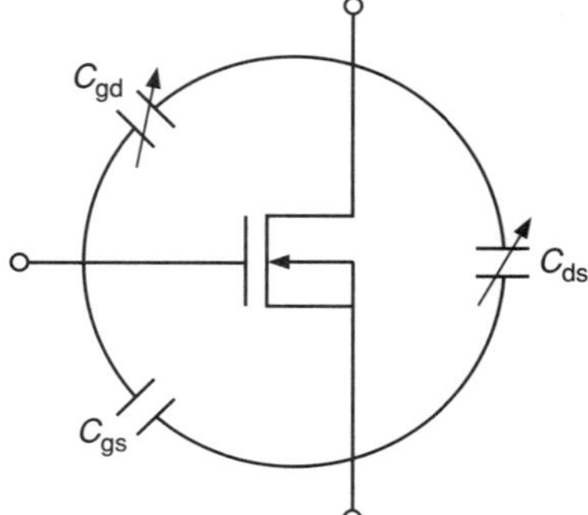

Figure 6.13 Terminal capacitances of power MOSFET.

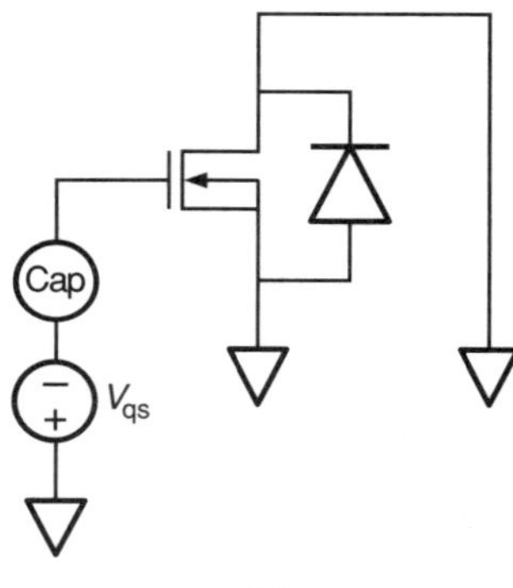

Figure 6.14(*a*) C_{iss} capacitance test measurement setup.

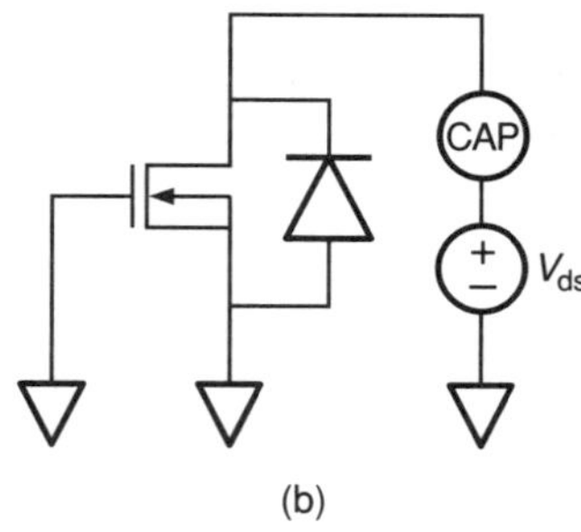

Figure 6.14(*b*) C_{oss} capacitance test measurement setup.

We can begin to determine the capacitance values of the model by subtracting Eq. 6.7 from Eq. 6.9, yielding the following relationship:

$$C_{iss} - C_{rss} = C_{gs} + C_{gd} - C_{gs} = C_{gs} \tag{6.10}$$

To determine C_{gs}, we simply measure the difference between the C_{iss} and C_{rss} at any point along the curve. This extraction is illustrated in Fig. 6.15. For our model, the gate-source capacitance will be

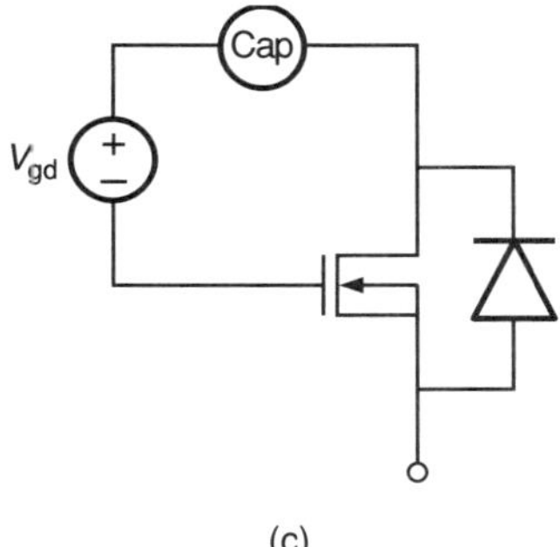

Figure 6.14(*c*) C_{rss} capacitance test measurement setup.

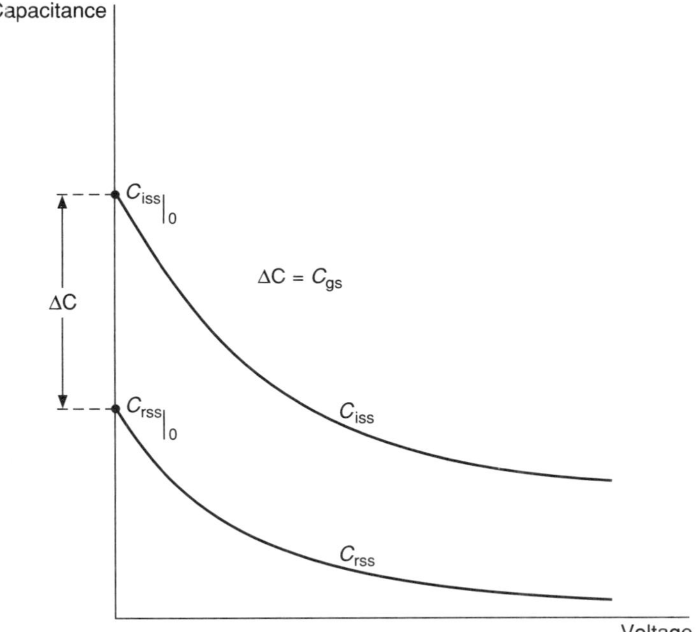

Figure 6.15 C_{gs} parameter extraction.

modeled with the capacitor, CGS. Set CGS equal to the measured value of gate-source capacitance.

With C_{gs} determined, we can turn our attention to C_{gd}. The C_{gd} capacitance is a highly nonlinear capacitance which will be modeled in our SPICE macromodel with three capacitors, two junction capacitances D_{MIN} and D_{MAX}, and a constant capacitance CGD. The constant capacitance CGD is determined from the gate-charge characteristics. CGD is determined by measuring the gate charge at which the gate

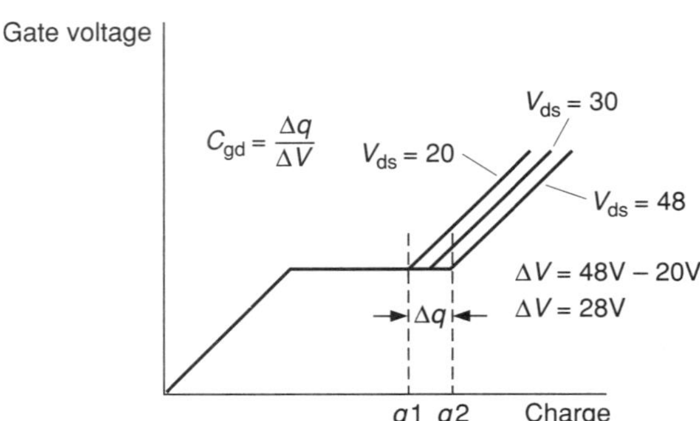

Figure 6.16 C_{gd} parameter extraction.

voltage begins to rise for different values of drain voltage. CGD is calculated from $\Delta q/\Delta V_{ds}$. Figure 6.16 illustrates this measurement. The value of CGD is determined from Eq. 6.11.

$$\text{CGD} = \frac{(\Delta q)}{(\Delta V_{ds})} \tag{6.11}$$

With the gate voltage much less than the drain voltage (saturated operation), the diode D_{MAX} is forward-biased and the diode D_{MIN} is reverse-biased. With D_{MIN} reverse-biased, the junction capacitance of D_{MIN} contributes to the gate-drain capacitance. The C_{rss} capacitance characteristic is essentially the capacitance characteristic of the reverse-biased D_{MIN} junction. To model the C_{rss} properly, we must extract the model parameters CJO, M, and VJ for the D_{MIN} diode from the C_{rss} curve. To determine CJO, M, and VJ, we will use the CJ_CALC program presented in Chap. 3 of this text.

CJ_CALC is a program which determines junction capacitance model parameters from measured capacitance-voltage characteristics. While learning to use the CJ_CALC program is beyond the scope of this chapter, the interested reader is directed to App. B.

Before the CJ_CALC program is run on the measured C_{rss} data, the measured capacitance (the capacitance entered into the CJ_CALC program) must be reduced by an amount equal to the value of CGD. Since our model places CGD in parallel with D_{MIN}, and since we have already calculated CGD, this constant capacitance should be subtracted from the data used in the CJ_CALC program. Enter the C_{rss} data after subtracting the CGD capacitance shown in Eq. 6.12. Often, a spreadsheet program is quite valuable in performing this calculation.

$$\text{For } n = 1 \text{ to the number of } C_{rss} \text{ data, } \text{CJ}_{(n)} = \text{C}_{rss(n)} - \text{CGD} \tag{6.12}$$

The final gate-drain capacitance term is the CJO of diode D_{MAX}. When in saturation, the drain-source voltage falls to almost zero and the gate-drain voltage forward-biases D_{MIN} and reverse-biases D_{MAX}. The value of CJO for D_{MAX} is determined experimentally with help from the GATE.BAT program. Start by setting CJO for diode D_{MAX} to 1000 pf, then simulate the gate-charge characteristics with the GATE.BAT program. Changing CJO for D_{MAX} will reposition the point where the gate voltage begins increasing again on the gate-charge characteristics plot. Adjust the CJO value (starting in 100- or 50-pf increments) until the simulated gate-charge characteristics match either the measured gate characteristics or the gate characteristics of the data sheet.

The C_{oss} capacitive curve illustrates the output characteristics of the power MOSFET. The C_{oss} curve includes the drain-source capacitance and the gate-drain capacitance (C_{rss}). The drain-source capacitance component is dominated by the junction capacitance of the body diode (D_{BODY}). To determine the capacitance parameters of D_{BODY} (CJO, M, and VJ), we must subtract the C_{rss} capacitance values (Eq. 6.9) from the C_{oss} values (Eq. 6.8). Subtracting these two sets of data leaves us with the capacitance-voltage characteristics of the body diode. A spreadsheet program can be extremely helpful in subtracting these two sets of data. Once these two values have been subtracted, the resulting data set, with help from the CJ_CALC program, can be used to calculate the capacitance model parameters of the body diode.

Parasitic elements

The power MOSFET macromodel includes four parasitic elements L_d, L_s, L_g, and R_g. The lead inductances will vary from package type to package type. For transistors in a TO-220 (plastic) package L_s and L_g are usually around 6 nH. The gate inductance is usually slightly less (4 nH is typical for TO-220). The gate resistance varies from device to device but is typically between 3 and 5 ohms. Table 6.4 illustrates several different package types and values for LD, LS, LG, and RG.

TABLE 6.4 Lead Inductance Values for Common Package Types

Package type	L_d	L_s	L_g	R_g
TO-220 (plastic)	4 nH	6 nH	6 nH	5 Ω
TO-204 (TO-3) (metal)	9 nH	12 nH	12 nH	5 Ω
TO-218 (plastic)	7.5 nH	10 nH	10 nH	5 Ω

Power MOSFET Model Extraction Demonstration

To demonstrate our modeling techniques, we will extract a model parameter from a Motorola MTD5N06 power MOSFET. Figure 6.17(*a*), (*b*), and (*c*) illustrate some of the data sheet information available for this transistor.

DC model parameters

We will begin our extraction with the DC characteristics of the MOSFET transistor core. To obtain KP, VTO, RS, and RD we need the linear characteristics of the MOSFET. Since this information is absent from the data sheet, the power MOSFET will be placed on a parametric analyzer as shown in Figs. 6.18(*a*) and (*b*). The result from the parametric analyzer was reformatted for RGRAPH, and the RGRAPH plot of this data is shown in Fig. 6.19. To calculate KP and VTO, the I-V data for VGS = 0.8 and higher was entered into the LINREG pro-

MAXIMUM RATINGS

Rating	Symbol	MTD5N05	MTD5N06	Unit
Drain-Source Voltage	V_{DSS}	50	60	Vdc
Drain-Gate Voltage (R_{GS} = 1 MΩ)	V_{DGR}	50	60	Vdc
Gate-Source Voltage — Continuous — Non-repetitive ($t_p \leq 50\ \mu s$)	V_{GS} V_{GSM}	±20 ±40		Vdc Vpk
Drain Current — Continuous — Pulsed	I_D I_{DM}	5 14		Adc
Total Power Dissipation @ T_C = 25°C Derate above 25°C	P_D	20 0.16		Watts W/°C
Total Power Dissipation @ T_A = 25°C Derate above 25°C	P_D	1.25 0.01		Watts W/°C
Total Power Dissipation @ T_A = 25°C (1) Derate above 25°C	P_D	1.75 0.014		Watts W/°C
Operating and Storage Junction Temperature Range	T_J, T_{stg}	−65 to +150		°C

THERMAL CHARACTERISTICS

Characteristic	Symbol	Value	Unit
Thermal Resistance — Junction to Case	$R_{\theta JC}$	6.25	°C/W
— Junction to Ambient — Junction to Ambient (1)	$R_{\theta JA}$	100 71.4	°C/W

ELECTRICAL CHARACTERISTICS (T_C = 25°C unless otherwise noted)

Characteristic	Symbol	Min	Max	Unit
OFF CHARACTERISTICS				
Drain-Source Breakdown Voltage (V_{GS} = 0, I_D = 0.25 mA) MTD5N05 MTD5N06	$V_{(BR)DSS}$	50 60	— —	Vdc
Zero Gate Voltage Drain Current (V_{DS} = 0.8 Rated V_{DSS}, V_{GS} = 0) T_J = 125°C	I_{DSS}	— —	0.2 1	mAdc

Figure 6.17(*a*) Maximum ratings per power MOSFET. (*Copyright of Motorola. Used by permission.*)

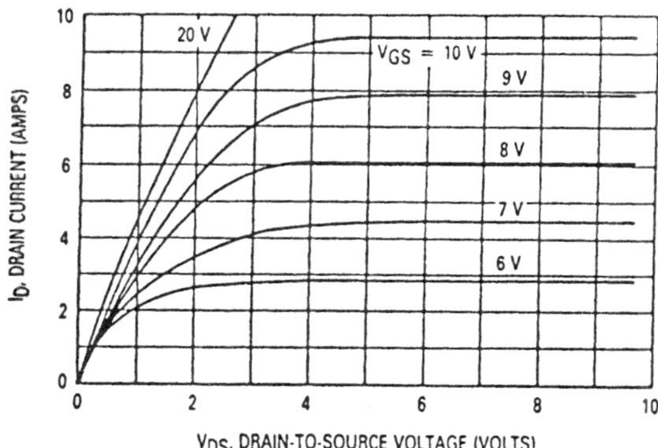

Figure 6.17(*b*) Family of curves. (*Copyright of Motorola. Used by permission.*)

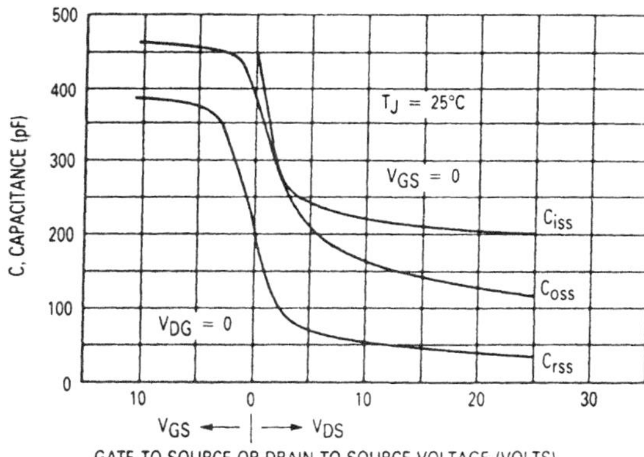

Figure 6.17(*c*) Capacitance variation of power MOSFET. (*Copyright of Motorola. Used by permission.*)

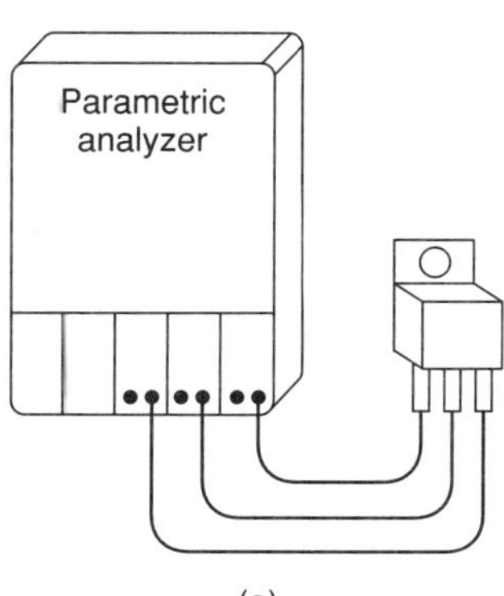

Figure 6.18(*a*) Test setup for linear characteristics test.

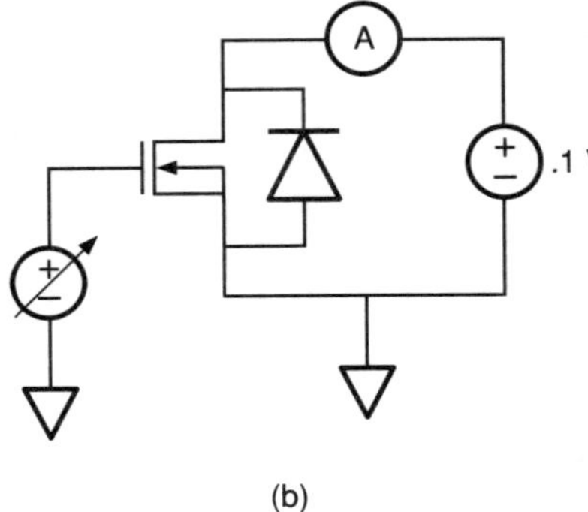

Figure 6.18(*b*) Measurement setup for linear characteristics test. (*Copyright of Motorola. Used by permission.*)

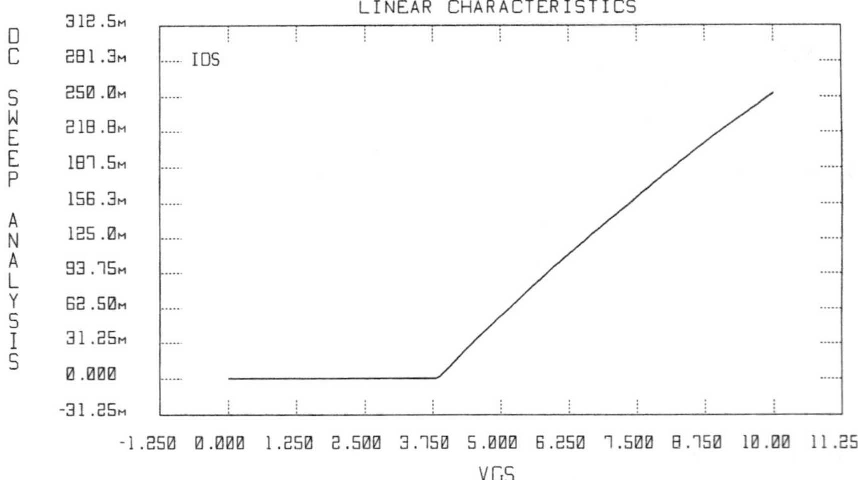

Figure 6.19 Measured linear characteristics of Motorola power MOSFET. (*Copyright of Motorola. Used by permission.*)

gram. The LINREG input and output data is shown in Figs. 6.20(*a*) and (*b*).

From the LINREG output, the x-axis intercept may be computed from Eq. 6.13.

$$x\text{-axis (intercept)} = -\frac{a}{b} = \frac{0.189}{0.049} = 3.85 \tag{6.13}$$

Equation 6.4 yields a value for VTO as shown in Eq. 6.14.

```
4                                   ! Number of data
3.900E+00       2.496E-03           ! X and Y values (Vgs, Ids)
4.000E+00       7.461E-03
4.100E+00       1.239E-02
4.200E+00       1.729E-02
```

Figure 6.20(*a*) LINREG (LRB) input file.

```
* * * * * * * * * * LINREG Results * * * * * * * *
*
* For the equation Y = a + b*X
*
* a    =   -0.18980E+00
* b    =    0.49311E-01
* chi2 = 0.000
*
```

Figure 6.20(*b*) LINREG (LRB) output file.

$$\text{VTO} = V_{\text{gs(intercept)}} - \frac{V_{\text{ds}}}{2} = 3.85 - \frac{0.1}{2} = 3.8 \text{ V} \tag{6.14}$$

Equation 6.5 yields a value for KP as shown in Eq. 6.15.

$$\text{KP} = \frac{\text{slope}}{V_{\text{ds}}} = \frac{0.049}{0.1} = 0.49 \text{ A/V}^2$$

To calculate RS and RD, the low-V_{gs} slope is extended to V_{gs} equal 10 V as shown in Fig. 6.21. The values of $I_{\text{d(high)}}$ and $I_{\text{d(low)}}$ were entered into the RS_CALC program with the values of KP, VTO, V_{gs}, and V_{ds}. RS_CALC produces the output shown in Fig. 6.22.

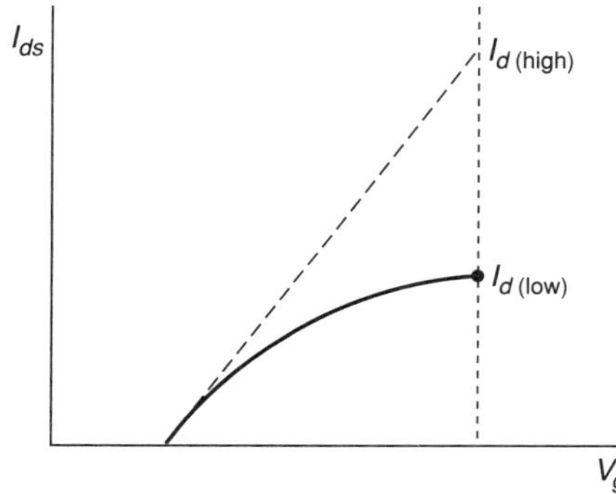

Figure 6.21 Linear MOSFET characteristics.

```
* * * * * * * * * * RS_CALC Results * * * * * * * * *
*
* RD = RS =     3.48965E-02
*
*
* * * * * * * * * * * * * * * * * * * * * * * * * * *
```

Figure 6.22 RS_CALC output.

TABLE 6.5 DC MOSFET Transistor Parameters

VTO	3.8 V
KP	0.49 A/V^2
RS	34.9 mΩ
RD	34.9 mΩ

Table 6.5 illustrates the DC parameters calculated for the core MOSFET model.

Figure 6.23 compares the measured and simulated linear characteristics. Figure 6.24 compares the measured and simulated family of curves. Both show an acceptably low 3 percent rms error.

The DC I-V relationship of the body diode is modeled with the parameters IS, RS, and N. To extract these parameters, we need the forward I-V characteristics of the diode. The transistor was connected to our parametric analyzer as shown in Fig. 6.25. By applying a

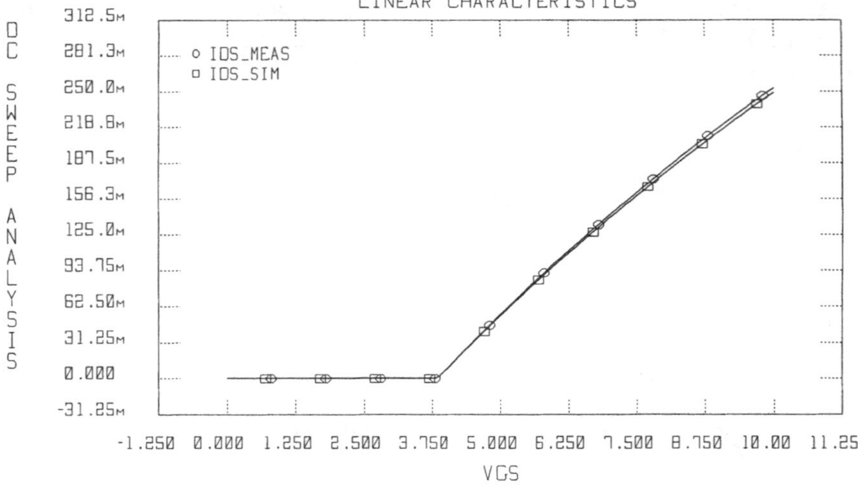

Figure 6.23 Measured and simulated linear characteristics.

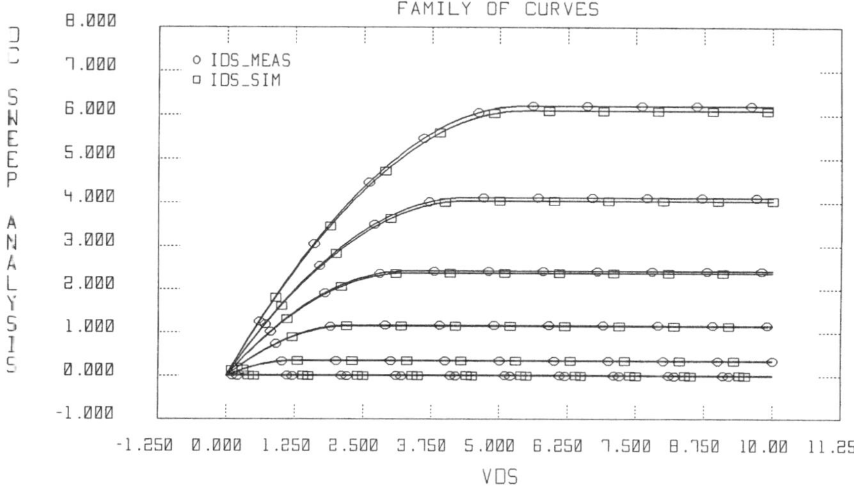

Figure 6.24 Measured and simulated family of curves.

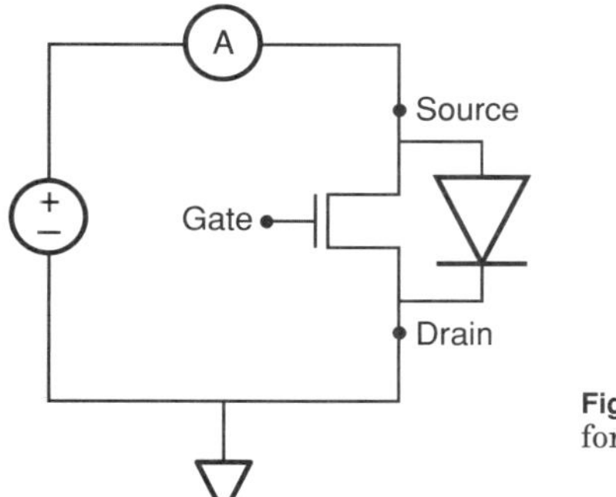

Figure 6.25 Measurement setup for body diode characteristics.

reverse bias to the drain-source terminals, we forward-bias the body diode. The measured I-V characteristics are shown in Fig. 6.26.

To determine the body diode parameters IS, RS, and N, we will follow the procedures outlined in Chap. 3. For those of you who have not read this section yet, a review of this chapter is necessary for understanding much of the remaining power MOSFET model.

All of the DC parameters of the body diode may be extracted from Fig. 6.26. The first parameter to be extracted is IS. IS is extracted from the *y*-axis intercept of the curve. The slope of the diode characteristics is used to determine N, and RS is determined from the nonlinear region at the top of the curve. These extractions are shown in Fig. 6.27, the extracted values are illustrated in Table 6.6 and the measured and simulated characteristics in Fig. 6.28.

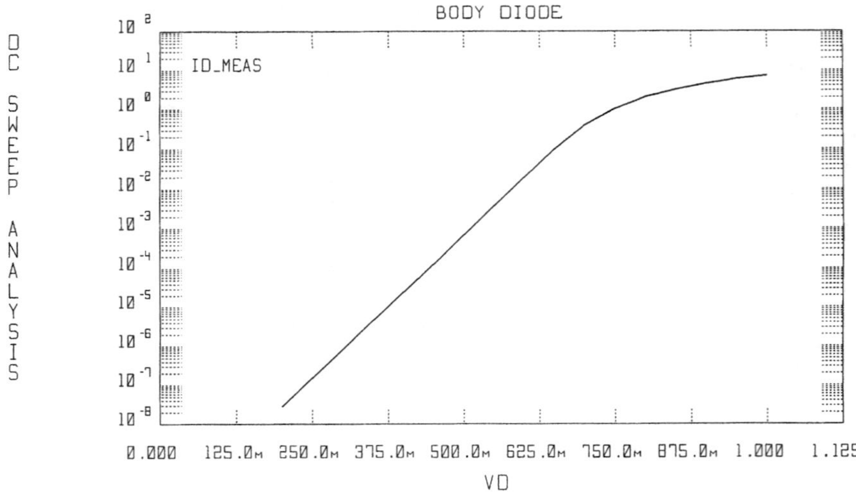

Figure 6.26 Measured body diode I-V characteristic.

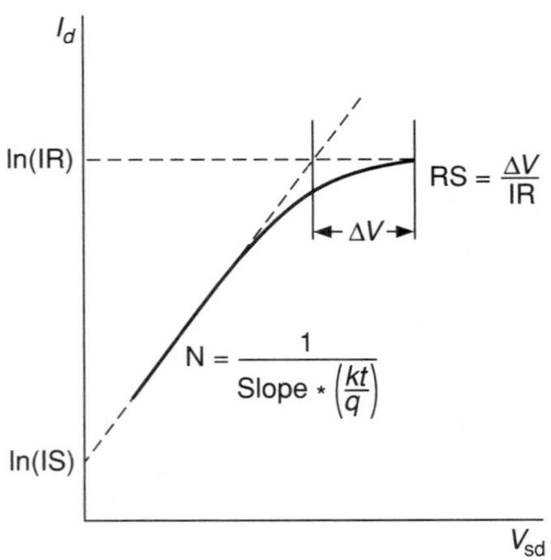

Figure 6.27 Extracting IS, RS, and N for the MOSFET body diode.

TABLE 6.6
Extracted Body Diode Parameters

IS	34.5 pA
RS	31 mΩ
N	1.15

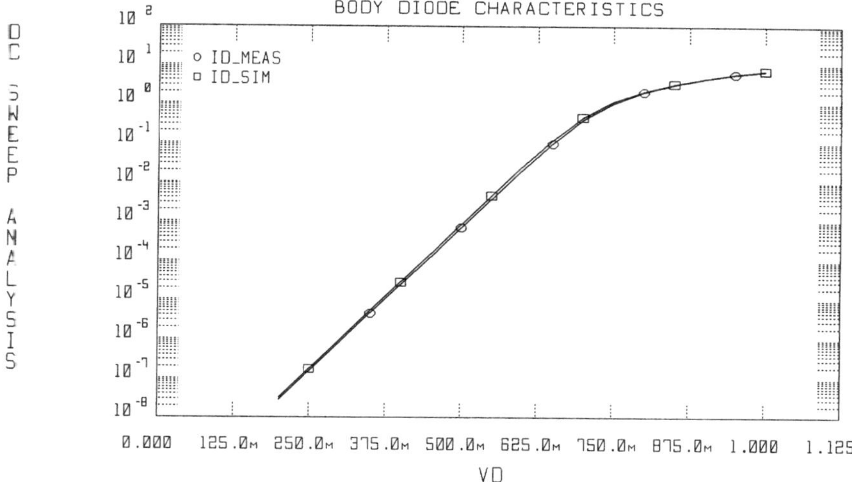

Figure 6.28 Measured and simulated body diode characteristics.

Capacitance parameters

To extract the capacitance parameters, we need the C_{oss}, C_{rss}, and C_{iss} characteristics of the transistor. While these curves are readily available in the data sheet information, we will measure these characteristics of the transistor and perform our extractions from the measured characteristics.

To measure the characteristics, an HP capacitance meter was connected to the transistor as shown in Fig. 6.14(*a*), (*b*), and (*c*). The measured characteristics are illustrated in Fig. 6.29.

We also need the gate-charge characteristics, and, although available on the data sheet, to ensure accuracy, the gate-charge characteristics will be measured. The measurement setup for this test is illustrated in Fig. 6.11(*b*) and the measured result is shown in Fig. 6.30.

We will begin the extraction by measuring the zero-biased values of C_{iss} and C_{rss}. From Eq. 6.10, the value of the gate-source capacitance CGS is given by Eq. 6.16.

$$\text{CGS} = C_{iss}(0) - C_{rss}(0) = 353.6 \text{ pf} - 174 \text{ pf} = 179.6 \text{ pf} \quad (6.16)$$

With the gate-source capacitance determined, we can turn our attention to the gate-drain capacitance. Of the three capacitors which model CGD in our macromodel (D_{MIN}, D_{MAX}, and CGD), CGD is the easiest to calculate. To calculate CGD, we need to measure the difference in gate charge where the V_{gs} voltage begins to rise after the plateau. This measurement should be done for the lowest value of V_{ds}

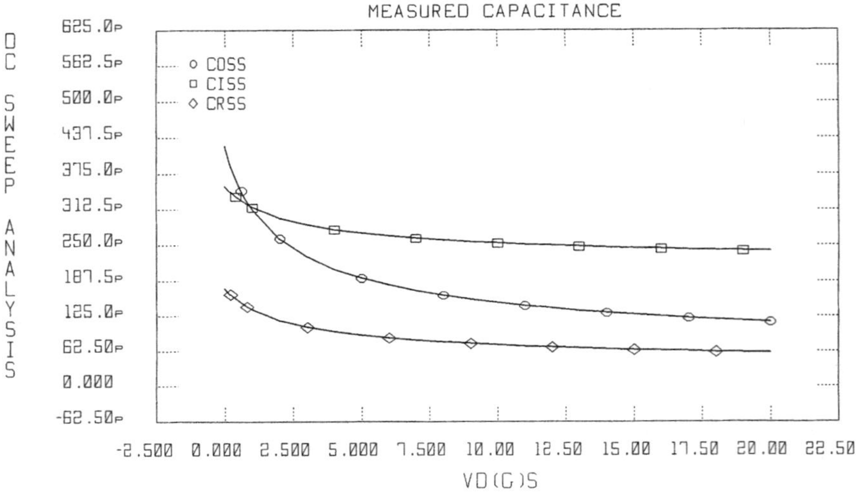

Figure 6.29 Measured power MOSFET capacitance characteristics.

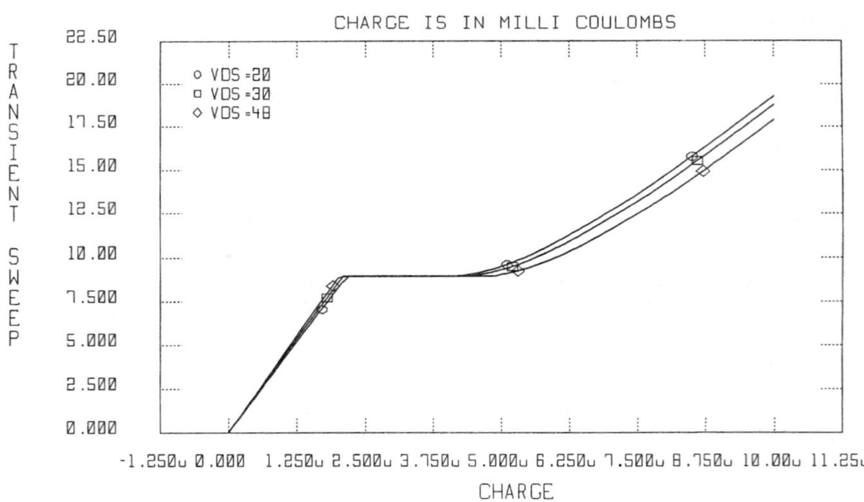

Figure 6.30 Measured gate-charge characteristics.

and the highest value of V_{ds}. This measurement is illustrated in Fig. 6.31. The difference in gate charge (Δq) and the difference in V_{ds} voltage (ΔV_{ds}) are then entered into Eqs. 6.11 and 6.17. We will set CGD to the value illustrated in Eq. 6.17.

$$\text{CGD} = \frac{(\Delta q)}{(\Delta V_{ds})} = \frac{0.8\ \text{nC}}{28\ \text{V}} = 28.5\ \text{pF} \tag{6.17}$$

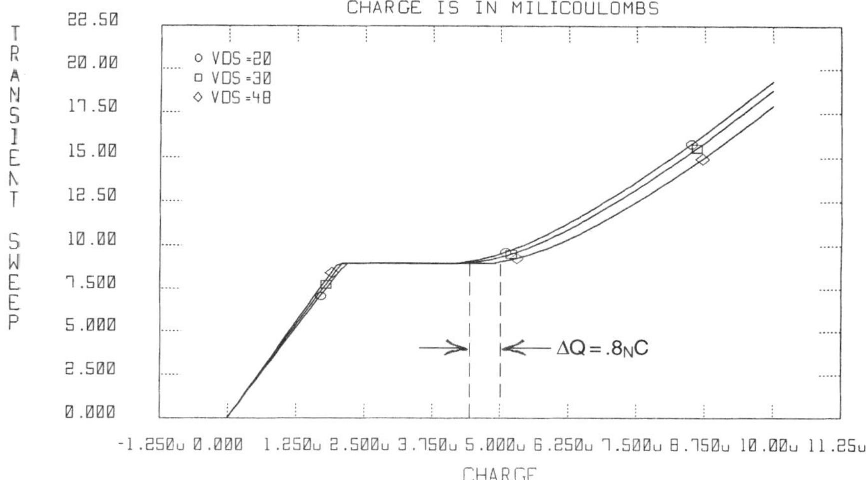

Figure 6.31 Measuring Δq between V_{ds} = 20 and 48 V.

Continuing with our gate-drain capacitance extractions, the characteristics of D_{MIN} must be determined. To determine CJO, M, and VJ for D_{MIN}, the capacitance-voltage characteristics of the gate-drain capacitance (C_{rss}) must be entered into CJ_CALC. *But, as prescribed by Eq. 6.12, before entering the measured C_{rss} values into CJ_CALC, the value of C_{rss} for each voltage value must be reduced by an amount equal to CGD.* (This modification of the C_{rss} values must be done to account for the fixed CGD value our model uses.)

After modifying the measured C_{rss} values, CJ_CALC can calculate the CJO, M, and VJ value of D_{MIN}. For our measured data, the C_{rss} values were entered into a spreadsheet program and 28.5 pF was subtracted from each C_{rss} data point. The resulting values were entered into the CJ_CALC program, and the output of the CJ_CALC program is shown in Table 6.7.

Simulating the C_{rss} curves and comparing these values with the C_{rss} data measured earlier demonstrates the accuracy of our model.

TABLE 6.7 Capacitance Parameters for D_{MIN} Calculated from CJ_CALC

CJO	145.5 pF
M	0.579
VJ	1.46 V

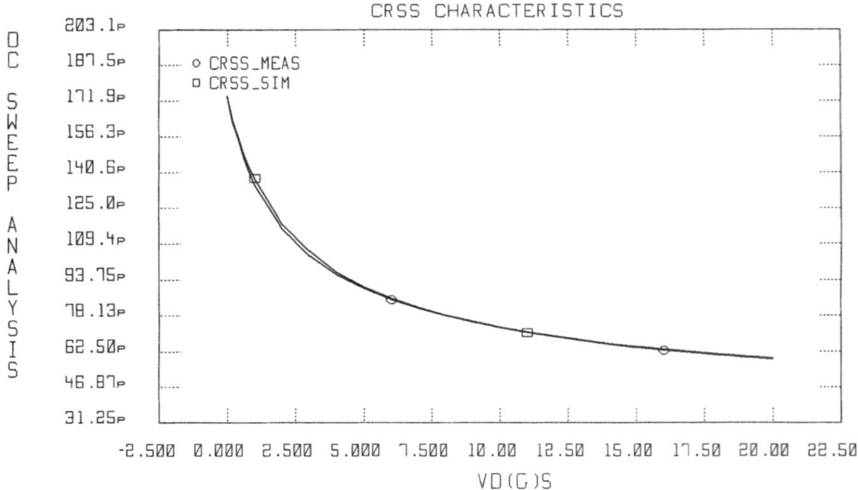

Figure 6.32 Measured and simulated C_{rss} characteristics.

Figure 6.32 illustrates the measured and simulated C_{rss} characteristics. The curves shown in Fig. 6.32 have an rms error of 0.72 percent!

The final component of the gate-drain capacitance is the zero-bias capacitance value (CJO) of D_{MAX}. The capacitance value of D_{MAX} determines the amount of charge on the gate at which the gate voltage begins to rise (after the plateau region). Increasing CJO increases the gate charge required to raise the gate voltage; decreasing CJO decreases the gate charge required to raise the gate voltage. Two different gate-charge characteristics are illustrated in Fig. 6.33.

The $CJO_{(DMAX)}$ value will be determined experimentally to produce the best fit of the gate-charge characteristics. Start by setting $CJO_{(DMAX)}$ to 1000 pF in the PMOS.MOD circuit file. Simulate the gate-charge characteristics with the GATE command file. Increase or decrease $CJO_{(DMAX)}$ to produce the best fit between the measured and simulated characteristics. Figure 6.34 illustrates the measured gate characteristics at $V_{ds} = 20$ V and simulations of the gate characteristics with $CJO_{(DMAX)}$ set to 1000, 900, and 800 pF. For our macromodel, a $CJO_{(DMAX)}$ set to 800 pF produces the best match to the measured characteristics. Figure 6.35 illustrates the measured and simulated gate-charge characteristics of our model.

Table 6.8 illustrates the gate-drain capacitance parameters which have been calculated to this point.

The final component of the capacitance model is the capacitance parameters of the body diode. The measured C_{oss} capacitance characteristics include both the body diode characteristics and the gate-

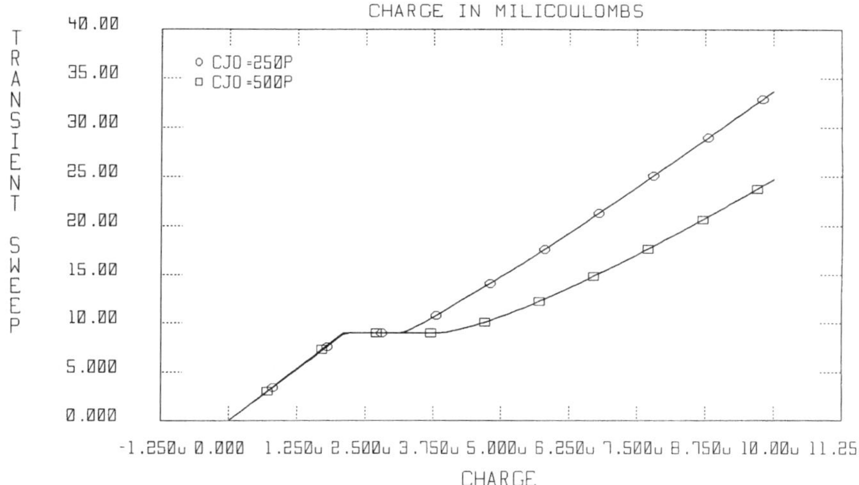

Figure 6.33 Gate-charge simulations for two different values of CJO (D_{MIN}).

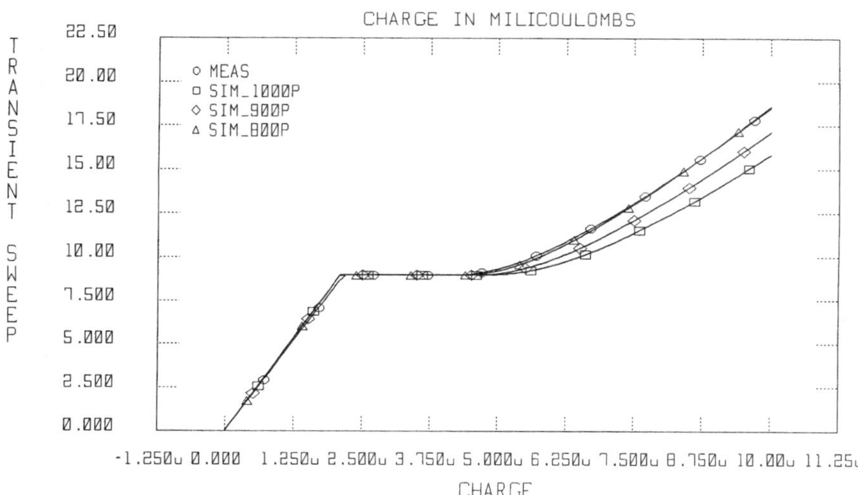

Figure 6.34 Simulated gate-charge characteristics with CJO set to 1000, 900, and 800 pF.

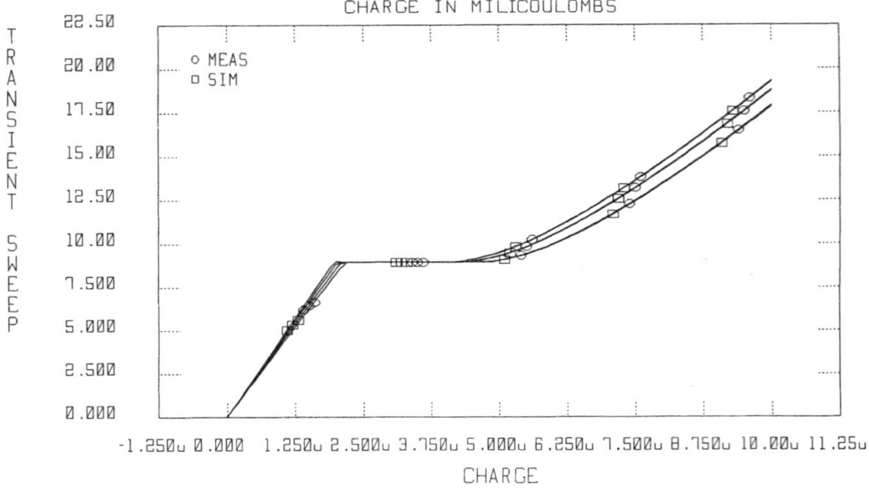

Figure 6.35 Simulated and measured gate-charge characteristics.

drain characteristics. To model the body diode, the gate-drain characteristics (C_{rss}) must be subtracted from the C_{oss} characteristics. Again, a spreadsheet program can be extremely useful in subtracting these values. Once the subtraction has been performed, the resulting data set is the capacitance-voltage characteristics of just the body diode.

To obtain the capacitance model parameters of the body diode, both the C_{oss} data set and the C_{rss} data set were entered into a spreadsheet program. At each value of C_{oss}, the C_{rss} value at that voltage was subtracted as shown in Fig. 6.36. The resulting data set was entered into the CJ_CALC program. The values shown in Table 6.9 illustrate the capacitance model parameters of the body diode calculated from CJ_CALC.

Figure 6.37 compares the measured and simulated C_{oss} characteristics of the power MOSFET model.

The values in Tables 6.8 and 6.9 and the value for CGS were entered into the PMOS.MOD file. The simulated values and the measured capacitance values of C_{oss}, C_{iss}, and C_{rss} are compared in Fig. 6.38. Figure 6.38 demonstrates a total rms error of 3.38 percent for the capacitance model, which is well within our limits of accuracy.

Both the I-V and capacitance model of our power MOSFET are within the 5 percent rms accuracy tolerance specified earlier. Practical simulation models can be built from this and other power MOSFETs by using the subcircuit introduced here.

TABLE 6.8 Capacitance Parameters for Gate-Drain Capacitance

CGD	28.5 pF	
CJO	325 pF	(for D_{MIN})
M	0.105	(for D_{MIN})
VJ	0.398 V	(for D_{MIN})
CJO	800 pF	(for D_{MAX})

Vcap	Coss	Crss	Coss-Crss
0	4.24E-10	1.74E-10	2.5E-10
0.2	3.91E-10	1.63E-10	2.28E-10
0.4	3.66E-10	1.55E-10	2.11E-10
0.6	3.45E-10	1.47E-10	1.98E-10
0.8	3.27E-10	1.41E-10	1.86E-10
1	3.12E-10	1.35E-10	1.77E-10
2	2.61E-10	1.16E-10	1.45E-10
3	2.29E-10	1.04E-10	1.25E-10
4	2.08E-10	9.6E-11	1.12E-10
5	1.92E-10	8.98E-11	1.02E-10
6	1.8E-10	8.5E-11	9.45E-11
7	1.7E-10	8.11E-11	8.84E-11
8	1.61E-10	7.78E-11	8.34E-11
9	1.54E-10	7.5E-11	7.91E-11
10	1.48E-10	7.26E-11	7.54E-11
11	1.43E-10	7.05E-11	7.22E-11
12	1.38E-10	6.87E-11	6.94E-11
13	1.34E-10	6.7E-11	6.69E-11
14	1.3E-10	6.55E-11	6.46E-11
15	1.27E-10	6.41E-11	6.26E-11
16	1.24E-10	6.29E-11	6.07E-11
17	1.21E-10	6.18E-11	5.9E-11
18	1.18E-10	6.07E-11	5.75E-11
19	1.16E-10	5.98E-11	5.59E-11
20	1.14E-10	5.89E-11	5.46E-11

Figure 6.36 Subtracting C_{rss} data from C_{oss} capacitance values. The final column of data ($C_{oss} - C_{rss}$) was entered into CJ_CALC.

TABLE 6.9 Capacitance Parameters for Body Diode

CJO	255 pF	(for D_{BODY})
M	0.433	(for D_{BODY})
VJ	0.371 V	(for D_{BODY})

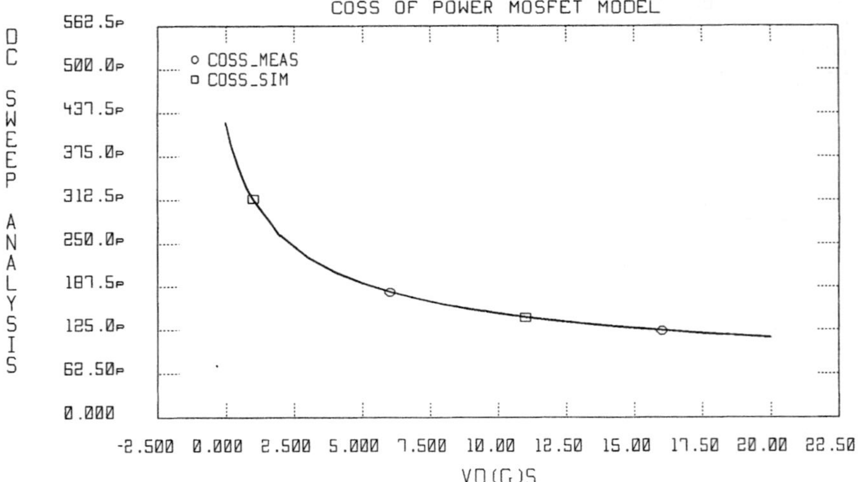

Figure 6.37 Measured and simulated C_{oss} characteristics of power MOSFET model.

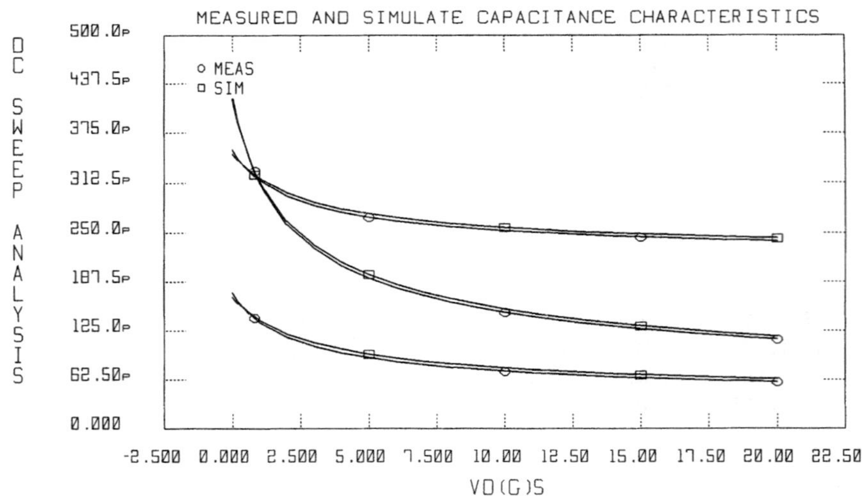

Figure 6.38 Simulated and measured capacitance characteristics of power MOSFET.

References

1. H. Nienhaus and J. Bowers, "A High Power MOSFET Model," *Powerconversion International,* January 1982.
2. J. Handcock, "A MOSFET Simulation Mode for Use in Microcomputer SPICE Circuit Analysis," *Proceedings of PCI,* 1987.
3. H. Yee and P. Lauritzen, "SPICE Models for Power MOSFETs: An Update," *Proceeding of IEEE APEC,* March 1988.
4. A. Vladimirescu, "A Power MOSFET Macromodel for Circuit Simulation," *Proceeding of PCIM,* 1990.
5. International Rectifier Application Note AN-975A and AN-954, International Rectifier, 233 Kansas St., Elsegundo, CA.
6. D. Haslam, "Simulating Power MOSFETs with SPICE," *Proceeding of HFPC,* May 1990.
7. Ed Oxner, personal correspondence, December 31, 1993.

Appendix A

Using the TC_CALC Program

TC_CALC is a simple FORTRAN (yes, some of us still use FORTRAN) program which calculates the first- and second-order temperature coefficients for resistors, capacitors, and inductors. The program is easy to use and may be executed interactively or in a batch mode.

To use the program, you must uncompress the files on the disk which accompanies this text. There is an installation program on the disk to do this for you. The installation program will create an `RCGV34` directory with several subdirectories. The TC_CALC program is located in the `RCGV34\UTIL` subdirectory.

Before using the TC_CALC program, you need measured values of resistance, capacitance, or inductance vs. temperature. The TC_CALC program requires at least four measured values but can accept up to 100. *Please note:* The accuracy of the temperature coefficients computed by TC_CALC is not related to the number of measured points! You can obtain the same level of accuracy with four measured points as you can with 100.

With measured values in hand, you are ready to run TC_CALC. To begin the program, change into the UTIL subdirectory and type:

```
TC_CALC
```

TC_CALC will respond with:

```
Enter Device Type (R, L, or C) >
```

To this, you must respond with R, L, or C. The program continues with:

```
Enter the Number of Measured Temperatures >
```

To this, you must enter a number between 4 and 100. The program continues with:

```
Enter the Room Temperature Value (25 C) of this Element >
```

Enter the room temperature value of the element. The program continues with:

```
Enter measurement temperature #1 for the element (in degrees C) >
```

Enter the first temperature. The program continues with:

```
Enter the element value at (temperature #1) >
```

The program repeats the previous two steps for all the measured temperatures.

The program responds by displaying the temperature coefficients TC_1 and TC_2. In addition to displaying the temperature coefficients, the program also displays the entered temperature, the measured element value at that temperature, and the element value calculated with TC_1 and TC_2. An example of the output display is shown in Fig. A.1.

For large files, the output display may scroll off the screen. Because of this, the same output information is saved in a file named TC_CALC.OUT after every run.

To use the TC_1 and TC_2 temperature coefficients for a resistor in SPICE, simply enter the values on the resistor statement as shown in Fig. A.2. The units of TC_1 and TC_2 are 1/degrees C and 1/(degrees C)^2. These units correspond to the units used for the SPICE resistor coefficients. (Users not familiar with the coefficients in SPICE should consult your SPICE users guide.) For capacitors and inductors, the values of TC_1 and TC_2 are entered as voltage-source voltage values, as described in Chap. 2.

The TCB Batch Program

While the TC_CALC program is easy to use for a small number of measured temperatures, manually entering a large number of measured values can be both tedious and error prone. To reduce the chance for errors, the TCB program was created.

TCB is a batch program which reads the data required for TC_CALC from a file. The TCB file requires an input file of the form

TC_CALC Results		
TC1 = 8.89799E-06		
TC2 = 2.50081E-08		
Temperature	Measured value	Calculated value
−40.0	9.99442E-01	9.99527E-01
−30.0	9.99511E- 01	9.99586E-01
−20.0	9.99585E-01	9.99650E-01
−10.0	9.99664E-01	9.99719E-01
0.0	9.99748E-01	9.99793E-01
10.0	9.99837E-01	9.99872E-01
20.0	9.99931E-01	9.99956E-01
30.0	1.00003E + 00	1.00005E + 00
40.0	1.00013E + 00	1.00014E + 00
50.0	1.00024E + 00	1.00024E + 00
60.0	1.00036E + 00	1.00034E + 00
70.0	1.00048E + 00	1.00045E + 00
80.0	1.00060E + 00	1.00057E + 00
90.0	1.00073E + 00	1.00068E + 00
100.0	1.00086E + 00	1.00081E + 00
110.0	1.00100E + 00	1.00094E + 00
120.0	1.00115E + 00	1.00107E + 00
130.0	1.00129E + 00	1.00121E + 00
140.0	1.00145E + 00	1.00135E + 00

Figure A.1 The TC_CALC output display.

```
RLOAD 3 7 10K TC = 8.897E-6,2.500E-8
```

Figure A.2 Temperature-dependent resistor syntax for SPICE.

shown in Fig. A.3. An example of the input file required for the TCB program is stored in the UTIL subdirectory under the name TC_CALC.RAW.

The data in the TCB input file is the exact same data you manually enter when running the TC_CALC program in an interactive mode.

To run the TCB batch program, simply change into the UTIL subdirectory and type:

```
TCB input_file_name
```

The program will display the normal prompts while reading the required data from the input file. As with TC_CALC, TCB displays the results of the temperature coefficient extraction on the screen and in the TC_CALC.OUT file.

```
R                   ; Device type (resistor)
19                  ; Number of temperature measurements
 1.0000             ; Nominal resistance value at 27C
-4.000000E + 01     ; Temperature #1 (-40C)
 9.994422E-01       ; Resistance at -40
-3.000000E + 01     ; Temperature #2 (-30C)
 9.995112E-01       ; Resistance at -30
-2.000000E + 01               "
 9.995852E-01                 "
-1.000000E + 01               "
 9.996642E-01                 "
 0.000000E + 00               "
 9.997482E-01                 "
 1.000000E + 01               "
 9.998372E-01                 "
 2.000000E + 01               "
 9.999312E-01                 "
 3.000000E + 01               "
 1.000030E + 00               "
 4.000000E + 01               "
 1.000134E + 00               "
 5.000000E + 01               "
 1.000243E + 00               "
 6.000000E + 01               "
 1.000357E + 00               "
 7.000000E + 01               "
 1.000476E + 00               "
 8.000000E + 01               "
 1.000600E + 00               "
 9.000000E + 01               "
 1.000729E + 00               "
 1.000000E + 02               "
 1.000863E + 00               "
 1.100000E + 02               "
 1.001002E + 00               "
 1.200000E + 02               "
 1.001146E + 00               "
 1.300000E + 02               "
 1.001295E + 00               "
 1.400000E + 02               "
 1.001449E + 00               "
```

Figure A.3 The TCB batch input file.

Appendix

B

Using the CJ_CALC Program

CJ_CALC is a FORTRAN program which calculates the junction capacitance parameters CJO, M, and VJ for SPICE diodes. Since SPICE uses these same equations for bipolar transistors, JFETs, and MOSFETs, the program can also be used to calculate the corresponding model parameters for each of these devices. Table B.1 illustrates the model parameters which may be calculated with CJ_CALC.

To use the CJ_CALC program, you must measure the junction capacitance while changing the reverse-bias voltage across the junction. In the case of the diode, you reverse-bias the diode; in the case of a bipolar transistor, you reverse-bias the base-emitter, base-collector junctions; for the JFET, you reverse-bias the gate-drain and gate-source junctions; and for the MOSFET, you reverse-bias the drain-bulk and source-bulk junctions.

The number of measurement values should be between 5 and 100. Above 10 data points, the accuracy of the extracted model parameters increases only slightly with a larger number of samples. In most extractions, you will achieve the same accuracy from 10 data points as you will with 100.

The measured data must include the measured capacitance at zero volts bias (no bias applied). For best results, more data points should

TABLE B.1 Equivalent Diode Capacitance Parameters

Diodes	BJTs	JFETs	MOSFETs
CJO	CJE, CJC, CJS	CGD, CGS	CDB, CSB
VJ	VJE, VJC, VJS	PB	PB
M	MJE, MJC, MJS	N/A	M

TABLE B.2 Typical Voltage Values for Junction Capacitance Test

Measurement #	Junction voltage
1	0
2	−0.2
3	−0.4
4	−0.6
5	−0.8
6	−1
7	−1.5
8	−2
9	−2.5
10	−3
11	−3.5
12	−4
13	−4.5
14	−5
15	−5.5
16	−6
17	−6.5
18	−7
19	−7.5
20	−8

be taken in the range of low reverse-bias voltages. Typical measurement values are shown in Table B.2.

With measured values in hand, you are ready to run CJ_CALC. To begin the program, change into the UTIL subdirectory and type:

```
CJ_CALC
```

CJ_CALC will respond with:

```
Enter the number of measured values >
```

To this, you must enter a number between 3 and 100. The program continues with:

```
Enter the zero-bias junction capacitance >
```

Enter the zero-bias capacitance value. The program continues with:

```
Enter bias voltage #1 >
```

Enter the first bias voltage as a negative quantity. The program continues with:

```
Enter measured capacitance at bias #1 >
```

Enter the measured capacitance. The program continues with:

```
Enter bias voltage #2 >
```

The program repeats the previous two steps for all the measured capacitance values.

The program responds by displaying the three capacitance model parameters CJO, VJ, and M. An example of the output display is shown in Fig. B.1. This same information is saved in the file named CJ_CALC.OUT.

In addition to the output file, two additional files are created. The first is named CJ_CALC.DAT and contains both the measured capacitance-vs.-voltage values and the values predicted with the newly solved values of CJO, VJ, and M. The second is named CJ_CALC.MAC and is an RGRAPH macrofile which contains a set of instructions for RGRAPH. Together, these two files can be used to produce an RGRAPH display of the measured and simulated capacitance values.

To produce a plot from the CJ_CALC.DAT and .MAC files, at the DOS prompt type:

```
RGRAPH CJ_CALC.DAT CJ_CALC.MAC
```

To this RGRAPH will first plot the measured capacitance values as shown in Fig. B.2. After a short delay, the program will continue by adding the simulated capacitance curve to the original plot. A typical example is shown in Fig. B.3. Notice, the second plot title displays the measured values of CJO, VJ, and M.

To use the CJO, M, and VJ values in SPICE, simply enter the values on the .MODEL statement, as shown in Fig. B.4.

The CJB Batch Program

While the CJ_CALC program is easy to use for a small number of measured temperatures, manually entering a large number of measured values can be both tedious and error prone. To reduce the chance for errors, the CJB program was created.

CJB is a batch program which reads the data required for CJ_CALC from a file. The CJB file requires an input file of the form shown

```
CJ_CALC Results

CJO = 200.000E-12
VJ  =   2.999
M   =   0.999
```

Figure B.1 The CJ_CALC output display.

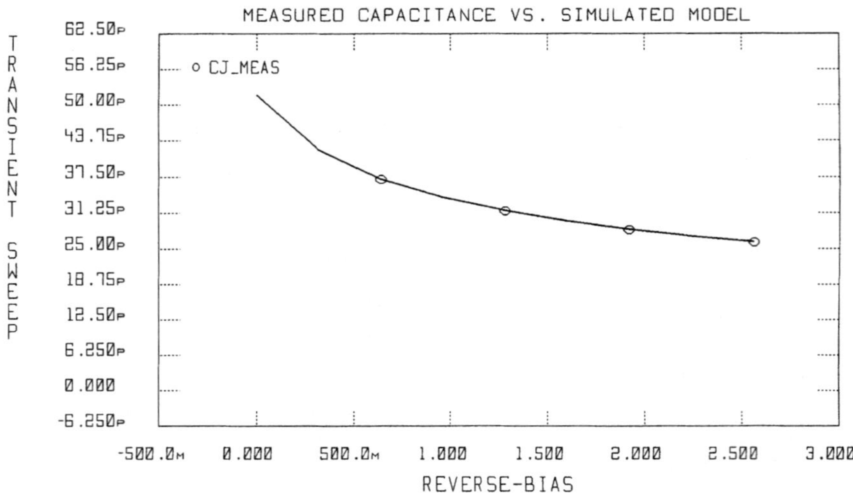

Figure B.2 Measured capacitance values.

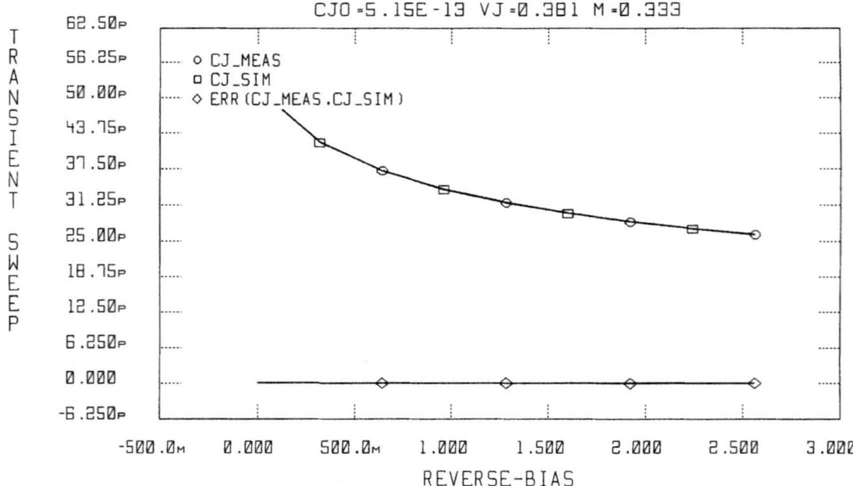

Figure B.3 Measured and simulated capacitance values.

```
.MODEL D1N4001 D (CJO = 1.2E-12 VJ = .394 M = .275...)
```

Figure B.4 Model statement syntax for SPICE.

```
   9            ; Number of capacitance measurements
  51.5e-12      ; Zero-bias capacitance value
   0.0          ; DC bias voltage #1 (Volts)
  51.6e-12      ; Measured capacitance at voltage #1
  −.32          ; DC bias voltage #2 (Volts)
 −42.1e-12      ; Measured capacitance at voltage #2
  −.64          "
  37.1e-12      "
  −.96          "
  33.9e-12      "
 −1.28          "
  31.6e-12      "
 −1.6           "
  29.8e-12      "
 −1.92
  28.3e-12
 −2.24          "
  27.1e-12      "
 −2.56          "
  26.1e-12      "
                ! Important: have a blank
                ! Line or two at the end of the
                ! Input file!
```

Figure B.5 The CJB batch input file:

in Fig. B.5. An example of the input file required for the CJB program is stored in the RCGV34 UTIL subdirectory under the name CJ_CALC.RAW.

The data in the CJB input file is the exact same data you manually enter when running the CJ_CALC program in an interactive mode.

To run the CJB batch program, simply change into the UTIL subdirectory and type:

```
CJB input_file_name
```

The program will display the normal prompts while reading the required data from the input file. As with CJ_CALC, CJB displays the results of the parameter extraction on the screen and in the CJ_CALC.OUT file. When complete, the CJB program invokes RGRAPH and produces an RGRAPH plot of the measured data.

To illustrate the CJB program, change into the UTIL subdirectory and type:

```
CJB CJ_CALC.RAW
```

Your output should match Figs. B.6 and B.7.

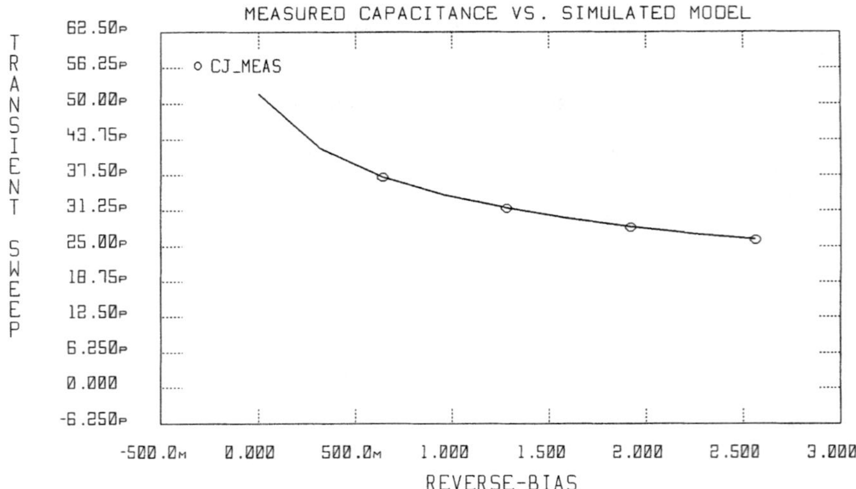

Figure B.6 Measured capacitance values from CJ_CALC.RAW.

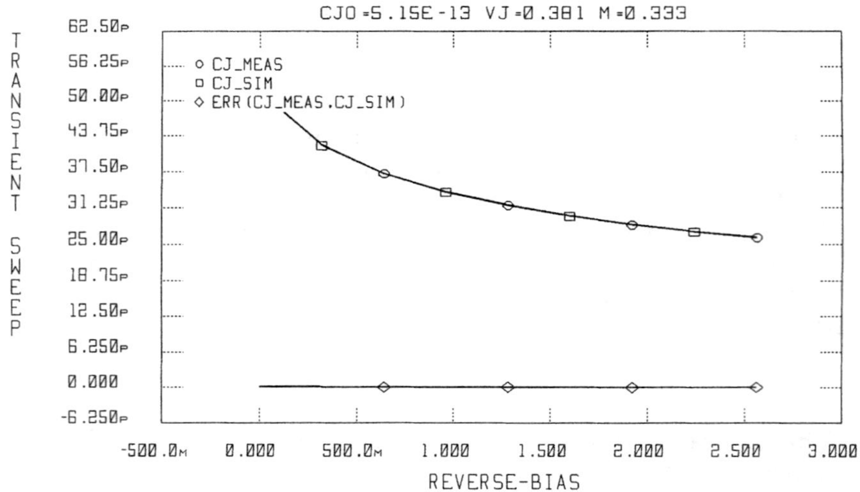

Figure B.7 Measured and simulated capacitance values from CJ_CALC.RAW

Appendix

C

Using the LINREG and LRB Programs

LINREG is another wonderful FORTRAN program which calculates the slope and intercept of a set of *x* and *y* data points. LINREG uses a linear least-squares method to predict the equation of a line which best fits the *x*, *y* data. Not only is this program extremely useful in device modeling, it's also an excellent general engineering tool. In this text, LINREG will be used to calculate model parameters of the diode, BJT, JFET, and MOSFET. Table C.1 illustrates the model parameters which may be calculated with LINREG.

To use the LINREG program, you must have a set of *x-y* data. The number of data points should be between 5 and 100. Typical LINREG data values are shown in Table C.2.

With the *x-y* data in hand, you are ready to run LINREG. To begin the program, change into the `RCGV34 UTIL` subdirectory and type:

```
LINREG
```

TABLE C.1 Model Parameters Which May Be Determined with LINREG.

Diodes	BJTs	JFETs	MOSFETs
IS	IS, ISE, ISC	IS	IS
N	NF,NR		
	NE, NC		
		VTO	VTO
		BETA	KP
		LAMBDA	LAMBDA

TABLE C.2 Typical LINREG *X-Y* Data

Measurement #	x	y
1	1	−5.3
2	2	7.1
3	3	19.7
4	4	31.7
5	5	43.6
6	6	55.2
7	7	67.3
8	8	79.3
9	9	91.6
10	10	103.3

LINREG will respond with:

```
Enter the number of X-Y values >
```

To this, you must enter a number between 3 and 100. The program continues with:

```
Enter the X and Y values for data point #1 >
```

Enter the x and y values (separated with either a space or a comma). The program repeats the previous step for all the x-y values.

The program responds by displaying the equation of a line ($y = a + b*x$) and the calculated values of a and b. In addition to the calculated slope and intercept, LINREG also prints the value of chi2. Chi2 is a measure of how close the data points fell to the line predicted by a and b. The smaller the value of chi2, the closer the data points are to the line. In the ideal case, where all the data points are on the line, chi2 is zero. An example of the output display is shown in Fig. C.1. This same information is saved in the file named LINREG.OUT.

In addition to the output file, two additional files are created. The first is named LINREG.DAT and contains both entered x and y data and predicted y values. The second is LINREG.MAC and is an RGRAPH MACRO file which contains a set of instructions for

LINREG Results

For the equation $y = a + b*x$

a = 0.360000E + 01
b = 0.57000E + 01
chi2 = 0.8300

Figure C.1 The LINREG output display.

RGRAPH. Together, these two files can be used to produce an RGRAPH display of the entered data and the resulting line.

To produce a plot from the LINREG.DAT and .MAC files, at the DOS prompt type:

```
RGRAPH LINREG.DAT LINREG.MAC
```

With these instructions, RGRAPH will plot the entered x and y data values. After a short delay, the program will continue by adding the plot of the predicted line. Examples of these plots are shown in Figs. C.2 and C.3. Notice that the second plot title displays the measured values of a and b.

The LRB Batch Program

While the LINREG program is easy to use for a small number of x-y data, manually entering a large number of values can be both tedious and error prone. To reduce the chance for errors, the LRB program was created.

LRB is a batch program which reads the data required for LINREG from a file. The LRB file requires an input file of the form shown in Fig. C.4. An example of the output file required for the LRB program is stored in the UTIL subdirectory under the name LINREG.RAW.

The data in the LRB input file is the exact same data you manually enter when running the LINREG program in an interactive mode.

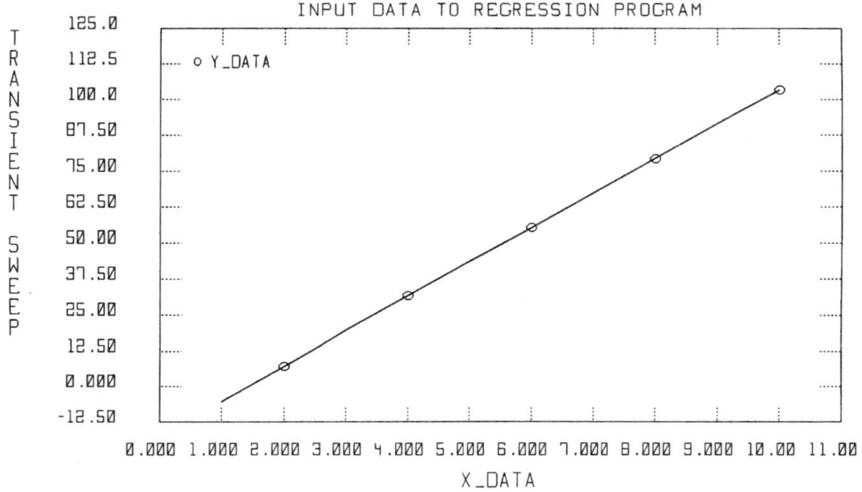

Figure C.2 RGRAPH plot from LINREG program.

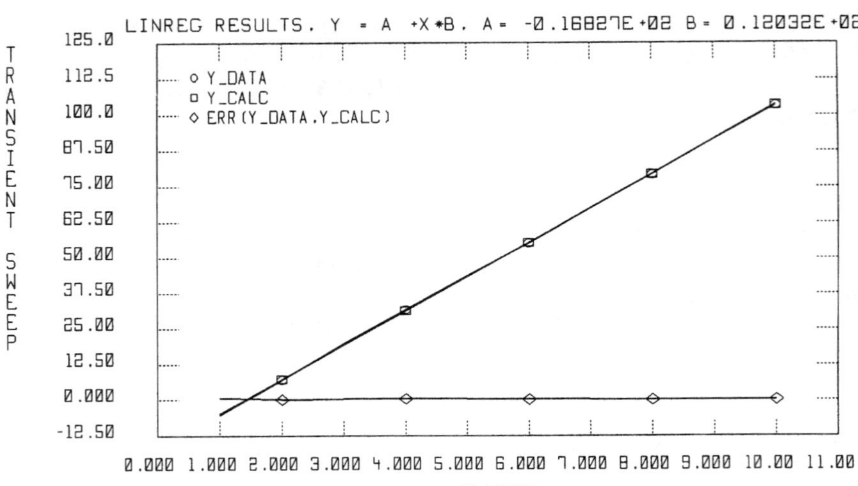

Figure C.3 RGRAPH plot from LINREG program.

```
10              ; Number of data points
 1     -5.3     ; X and Y #1
 2      7.1     ; X and Y #2
 3     19.7     ; "
 4     31.7
 5     43.6
 6     55.2
 7     67.3
 8     79.3
 9     91.6
10    103.3
                ! Important: have a blank
                ! line or two at the end of the
                ! input file!
```

Figure C.4 The LRB batch input file.

To run the LRB batch program, simply change into the UTIL subdirectory and type:

```
LRB input_file_name
```

The program will display the normal prompts while reading the required data from the input file. As with LINREG, LRB displays the results of the extraction on the screen and in the LINREG.OUT file. When complete, the LRB program invokes RGRAPH and produces an RGRAPH plot of the measured data.

To illustrate the LRB program, change into the UTIL subdirectory and type:

```
LRB LINREG.RAW
```

Your output should match Figs. C.5 and C.6.

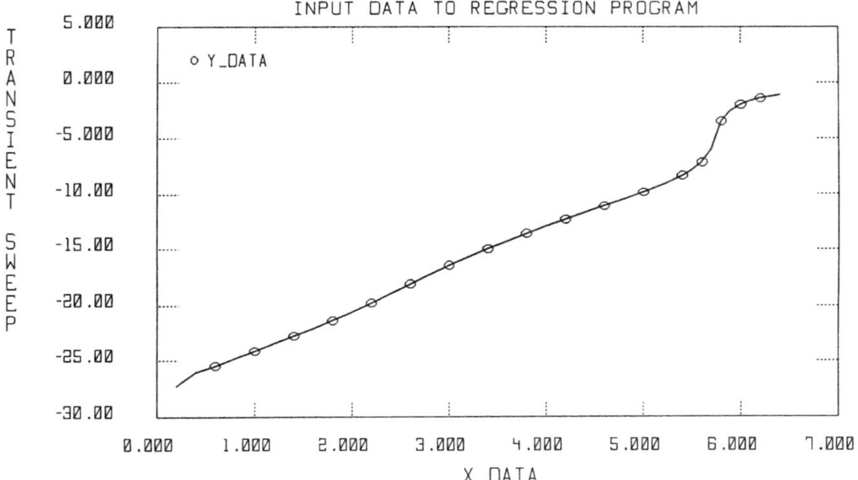

Figure C.5 RGRAPH plot from LRB program.

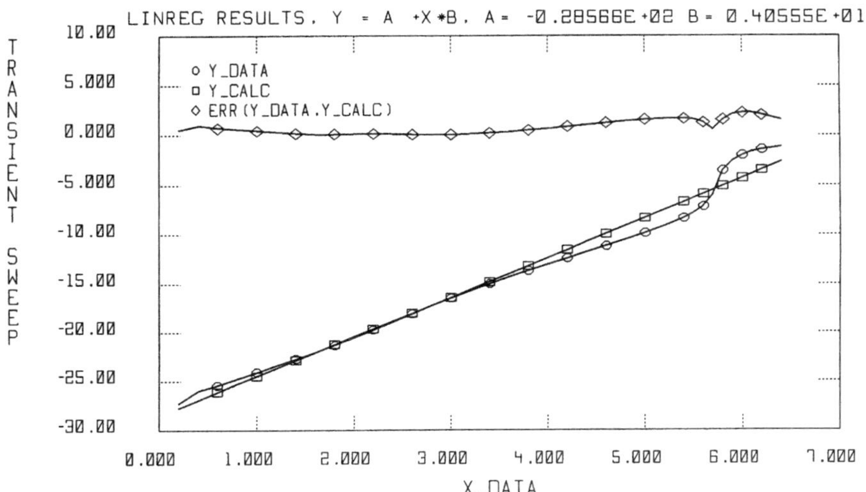

Figure C.6 RGRAPH plot from LRB program.

Appendix

D

Analog Behavioral Models

Analog behavioral modeling is a type of modeling which expresses voltages and current in terms of mathematical relationships rather than normal electrical relationships. For example, a resistor follows the classic electrical relationship:

$$V = I * R \tag{D.1}$$

This states simply that the voltage across the terminals of the resistor is equal to the product of the current and the resistance value. But a behavioral model might follow the following mathematical relationship:

$$V = \sqrt{T(V(1) * V(2))} \tag{D.2}$$

The ability to express voltages and currents in mathematical relationships rather than electrical relationships is the basic difference between traditional macromodeling (subcircuits) and behavioral modeling.

In this section, several different behavioral subcircuits will be introduced. The subcircuits perform mathematical operations on the voltages applied to the circuit. The output voltage will be mathematically related to the input voltage. All of the subcircuits you are about to see are stored on your disk in the `RCGV34\LIB` subdirectory and are yours to use.

The subcircuits which have been placed at the end of this Appendix are all based on standard SPICE elements and, as such, will work in any SPICE-based simulation software. Most of the subcircuits use what will be for many a new class of elements: the polynomial-dependent sources. The polynomial-dependent sources operate in much the same manner as standard controlled sources. For example, the standard voltage-controlled voltage source has the following syntax:

```
EXXXXX N+ N- (NC+,NC-) GAIN
```

This dependent generator is located between the nodes N + and N-. The voltage across the N + and N- terminals is equal to GAIN* (*V*(NC+) − *V*(NC-)). This represents the relationship of the standard SPICE voltage-controlled voltage source.[1] The polynomial-controlled source has a different syntax as shown in the following:

```
EXXXXX N+ N- POLY(X)...
```

The *X* is a variable you assign when you enter the element into SPICE. The variable *X* represents the number of control voltages which control the output of the dependent generator. That's right. This element may have an output voltage which depends on one or more controlling voltages![1]

If we make the following assignment in SPICE:

```
EXXXXX N+ N- POLY(1) (NC+, NC-) p0 p1 p2 p3...
```

SPICE determines that the output of this source will be determined by just one control voltage (the voltage across NC + and NC-). But instead of a simple multiple of the control voltage, the output voltage of this source will be determined by the value of the coefficients $p0$, $p1$, $p2$, etc. The output voltage of this source may now be expressed as:

$$V_{out} = p0 + p1 * V_c + p2 * V_c^2 + p3 * V_c^3 + \dots$$

where V_c is *V*(NC+, NC-).

The output voltage of this dependent generator is determined by the coefficients $p0$, $p1$, $p2$,..., and a general polynomial expansion (hence, the name "polynomial-dependent sources"). While this is useful if we need to make a subcircuit which generates the square or cube output of the input voltage, a circuit element of more interest is the polynomial source controlled by two voltages.

```
EXXXXX N+ N- POLY(2) (NC1+,NC1-) (NC2+,NC2-) p0 p1 p2 p3...
```

This generator is controlled by two independent voltages, and the output of this new source is equal to:

$$V_{out} = p0 + p1 * V1 + p2 * V2 + p3 * V1^2 + p4 * V1 * V2 + p5 * V2^2 + p6 * V1^2 * V2 \ldots$$

where $V1$ is V(NC1+, NC1-) and $V2$ is V(NC2+, NC2-).

To make a simple adder circuit, we need the polynomial-dependent source syntax shown in the following:

```
EXXXXX N + N- POLY(2) (NC1+, NC1-) (NC2+, NC2-) 0 1 1
```

In this expression, $p0$ is set to zero; $p1$ and $p2$ are both set to one. If we substitute these values into the previous expression for V_{out}, we arrive at:

$$V_{out} = 0 + V1 + V2$$

If we reverse the sign of the $p2$ coefficient, the adder becomes an ideal subtracter:

$$V_{out} = 0 + V1 - V2$$

If we set $p0$, $p1$, and $p2$, and $p3$ to zero and set $p4$ to one, the output voltage becomes the product of the voltages $V1$ and $V2$:

$$V_{out} = 0 + 0 + 0 + 0 + V1 * V2$$

In other words, an ideal multiplier circuit!

All of the behavioral models in this text will be constructed from polynomial-dependent sources. Some of the behavioral subcircuits will use a single polynomial-dependent source, while others will use two or more. Some of the circuits are relatively straightforward and easy to understand, while others will require the user write the complete set of nodal equations to relate the input and output voltages.

While a full explanation of the polynomial sources is beyond the scope of this text, this basic introduction should be all that is needed to understand most of the behavioral subcircuits which follow. Readers who are interested in more on polynomial-dependent sources are directed to their SPICE user's guides. Remember, these are standard SPICE elements, and any good SPICE-based simulator will be able to use these sources.

The behavioral subcircuits which are included in the `RCGV34\LIB` subdirectory are shown in Table D.1.

TABLE D.1 Behavioral Subcircuits

Function	No. of input voltages	Output units
Adder	2	Volts
Subtracter	2	Volts
Multiplier	2	Volts
Divider	2	Volts
Square	1	Volts
Square Root	1	Volts
Integrator	1	Volts
Differentiator	1	Volts
Natural Log	1	Volts
Exponential	1	Volts
Sine	1	Volts
Cosine	1	Volts
Voltage-controlled resistor	1	Ohms
Voltage-controlled capacitor	1	Farads
Voltage-controlled inductor	1	Henrys
Voltage-controlled potentiometer	2	Ohms

Adder Subcircuit

```
Advanced Techniques
*
* RCG Research, Inc. 3/13/90
*
* This is a three port adder circuit.
* The output voltage is Vin1 + Vin2
*
.SUBCKT ADD 1 2 3 4 5 6
*
* 1: Vin1+
* 2: Vin1-
* 3: Vin2+
* 4: Vin2-
* 5: Vout+
* 6: Vout-
*
R12 1 2 1000G
R34 3 4 1000G
ROUT 5 6 1000G
E1 5 6 POLY(2) (1,2) (3,4) 0 1 1
.ENDS
*
* Test the circuit with a voltage sweep
*
V1 1 0 SIN 0 2 5 .125 10
V2 2 0 SIN 0 1 2.5 .125 5
X1 1 0 2 0 3 0 ADD
R1 3 0 1000G
.TRAN .01 1
.PRINT TRAN V(1) V(3) V(2)
.END
```

Adder Subcircuit

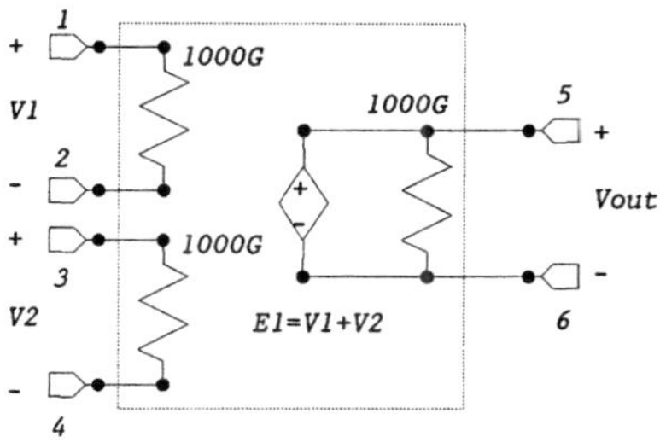

Subtraction Subcircuit

```
Advanced Techniques
*
* RCG Research, Inc. 3/13/90
*
* This is a three port adder circuit.
* The output voltage is Vin1 - Vin2
*
.SUBCKT SUB 1 2 3 4 5 6
*
* 1: Vin1+
* 2: Vin1-
* 3: Vin2+
* 4: Vin2-
* 5: Vout+
* 6: Vout-
*
R12 1 2 1000G
R34 3 4 1000G
ROUT 5 6 1000G
E1 5 6 POLY(2) (1,2) (3,4) 0 1 -1
.ENDS
*
* Test the circuit with a voltage sweep
*
V1 1 0 SIN 0 2 5 .125 10
V2 2 0 SIN 0 1 2.5 .125 5
X1 1 0 2 0 3 0 SUB
R1 3 0 1000G
.TRAN .01 1
.PRINT TRAN V(1) V(3) V(2)
.END
```

Subtractor Subcircuit

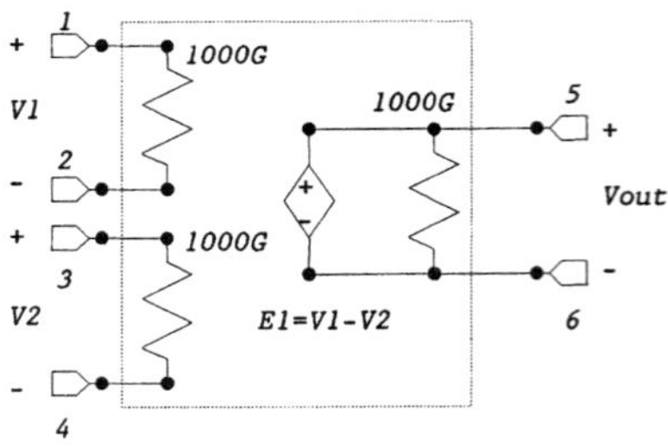

Multiplication Subcircuit

```
Advanced Techniques
*
* RCG Research, Inc. 3/13/90
*
* This is a three port multiplier circuit.
* The output voltage is Vin1 * Vin2
*
.SUBCKT MUL 1 2 3 4 5 6
*
* 1: Vin1+
* 2: Vin1-
* 3: Vin2+
* 4: Vin2-
* 5: Vout+
* 6: Vout-
*
R12 1 2 1000G
R34 3 4 1000G
E1 5 6 POLY(2) (1,2) (3,4) 0 0 0 0 1
.ENDS
*
* Test the circuit with a voltage sweep
*
V1 1 0 SIN 0 5 5 .125 10
V2 2 0 SIN 0 4 2.5 .125 5
X1 1 0 2 0 3 0 MUL
R1 3 0 1000G
.TRAN .01 1
.PRINT TRAN V(1) V(3) V(2)
.END
```

Multiplier Subcircuit

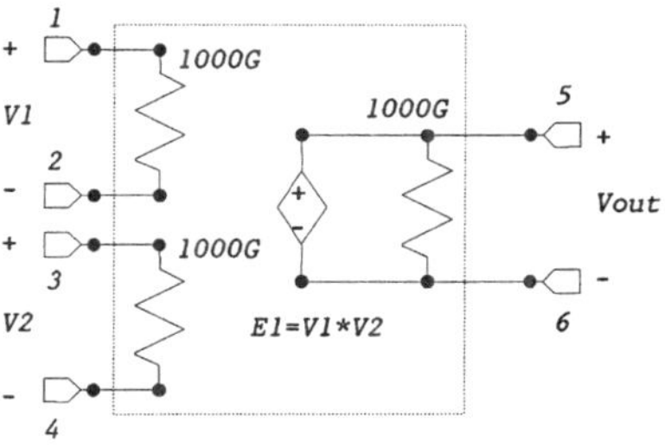

Divider Subcircuit

```
Advanced Techniques
*
* RCG Research, Inc. 3/13/90
*
* This is a three port divider circuit.
* The output voltage is Vnum / Vden.
*
.SUBCKT DIV 1 2 3 4 5 6
*
* 1: Vnum+
* 2: Vnum-
* 3: Vden+
* 4: Vden-
* 5: Vout+
* 6: Vout-
*
R12 1 2 1000G
R34 3 4 1000G
R70 7 0 1000G
R56 5 6 1000G
G1 0 7 1 2 1
G2 7 0 POLY(2) (3,4) (7,0) 0 0 0 0 1
E1 5 6 7 0 1
.ENDS
*
* Test the circuit with a voltage sweep
*
V1 1 0 SIN 0 4 5 .125 10
V2 2 0 SIN 0 4 2.5 .125 5
X1 1 0 2 0 3 0 DIV
R1 3 0 1000G
.TRAN .01 1
.PRINT TRAN V(1) V(3) V(2)
.OPTIONS ITL4=40 NOINTR
.END
```

Divider Subcircuit

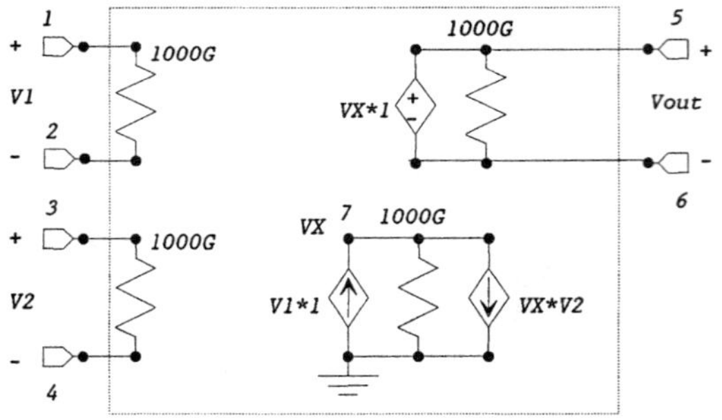

Square Subcircuit

```
Advanced Techniques
*
* RCG Research, Inc. 3/13/90
*
* This is a two port squaring circuit.
* The output voltage is Vin * Vin
*
.SUBCKT X2 1 2 3 4
*
* 1: Vin+
* 2: Vin-
* 3: Vout+
* 4: Vout-
*
R12 1 2 1000G
R34 3 4 1000G
ROUT 3 4 1000G
E1 3 4 POLY(1) (1,2) 0 0 1
.ENDS
*
* Test the circuit with a voltage sweep
*
V1 1 0 SIN 0 4 5 .125 10
X1 1 0 2 0 X2
R1 2 0 1000G
.TRAN .01 1
.PRINT TRAN V(1) V(2)
.END
```

X^2 Schematic and Block Diagram

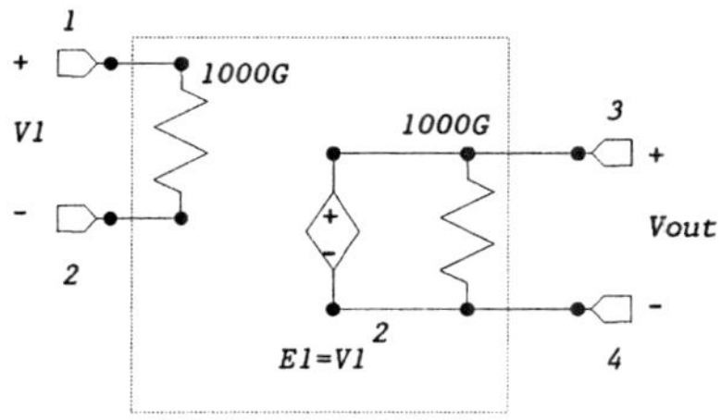

Square-Root Subcircuit

```
Advanced Techniques
*
* RCG Research, Inc. 3/13/90
*
* This is a square root circuit.
* The output voltage is the square root of the input voltage.
* This is a nested subcircuit and uses the DIV subcircuit in
* a feedback loop.
*
.SUBCKT SQRT 1 2 3 4
*
* 1: Vin+
* 2: Vin-
* 3: Vout+
* 4: Vout-
*
X1 1 2 3 4 3 4 DIV
.MODEL DMOD D
.INCLUDE DIV.MOD
.ENDS SQRT
*
* Test the circuit with a voltage sweep
*
V1 1 0 PWL 0V 0S 100V 100S
X1 1 0 2 0 SQRT
R1 2 0 1000G
.TRAN 1 100
.PRINT TRAN V(1) V(2)
.OPTIONS NOINTR LIMPTS=1MEG
.END
```

SQRT( ) Subcircuit

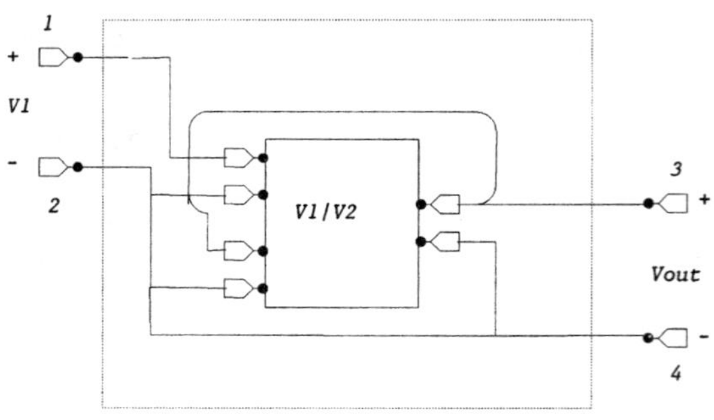

Integration Subcircuit

```
Advanced Techniques
*
* RCG Research, Inc. 3/13/90
*
* This is an integrator circuit.
* The output voltage is the time integral of Vin(t).
*
.SUBCKT INTGR8 1 2 3 4
*
* 1: Vin1+
* 2: Vin1-
* 3: Vout+
* 4: Vout-
*
G34 4 3 1 2 1
R12 1 2 1000G
C34 3 4 1
R34 3 4 1000G
.ENDS
*
* Test the circuit with a sine wave
*
VSIN 1 0 SIN (0 1 .15915 0 0)
* This initialization needed to avoid problems in finding the DC op. point
*.ic vout(0+) = -vin(0)
.IC V(2) = 0
X1 1 0 2 0 INTGR8
.TRAN .1 20
.PRINT TRAN V(1) V(2)
.END
```

INTGR8 Subcircuit

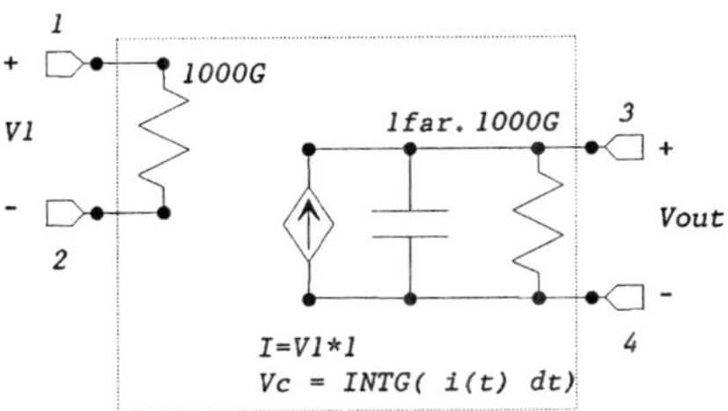

Differentiation Subcircuit

```
Advanced Techniques
*
* RCG Research, Inc. 3/13/90
*
* This is a differentiator circuit.
* The output voltage is dVin(t).
*
.SUBCKT DVDT 1 2 3 4
*
* 1: Vin1+
* 2: Vin1-
* 3: Vout+
* 4: Vout-
*
G34 4 3 1 2 1
R12 1 2 1000G
L34 3 4 1
R34 3 4 1000G
.ENDS DVDT
* LTE timestep control may produce funny results
.OPTIONS LVLTIM=1
*
* Test the circuit with a sine wave
*
VSIN 1 0 SIN (0 1 .15915 0 0)
X1 1 0 2 0 DVDT
.TRAN .1 20
.PRINT TRAN V(1) V(2)
.OPTIONS NOINTR
.END
```

DVDT Subcircuit

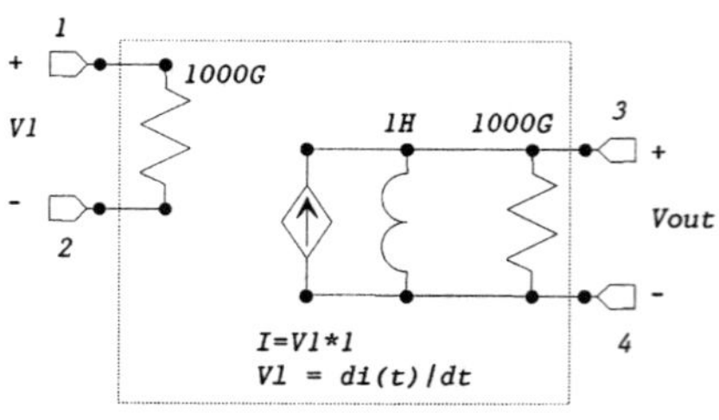

Exponential Subcircuit

```
Advanced Techniques
*
* RCG Research, Inc. 3/13/90
*
* This is a two port exponentiator circuit.
* The output voltage is the exponential of the input voltage.
*
.SUBCKT EXP 1 2 3 4
*
* 1: in+
* 2: in-
* 3: out+
* 4: out-
*
VX 0 5
E1 6 5 1 0 .0258665
H1 3 4 POLY(1) VX 1 1E12
R1 1 2 1000G
D1 6 0 DIODE
.MODEL DIODE D IS=1E-12
ROUT 3 4 1000G
.ENDS EXP
.OPTIONS GMIN=1E-18
*
* Test the circuit with a voltage sweep
*
VTEST 1 0
X1 1 0 2 0 EXP
.DC VTEST 1 10 .5
.PRINT DC V(1) V(2)
.END
```

EXP( ) Subcircuit

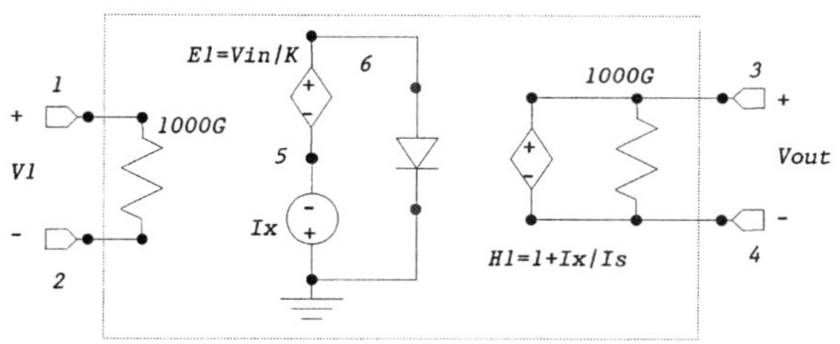

Natural Log Subcircuit

```
Advanced Techniques
*
* RCG Research, Inc. 3/13/90
*
* This is a two port natural logarithm circuit.
* The output voltage is the logarithm of the input voltage.
*
.SUBCKT LN 1 2 3 4
*
* 1: in+
* 2: in-
* 3: out+
* 4: out-
*
EOUT 3 4 5 0 38.66
R1 1 2 1000G
ROUT 3 4 1000G
G1 0 5 POLY(1) 1 0 -1 1
D1 5 0 DIODE
.MODEL DIODE D IS=1
.ENDS LN
.OPTIONS GMIN=1E-18
*
* Test the circuit with a voltage sweep
*
VTEST 1 0
X1 1 0 2 0 LN
.DC VTEST 1 10 .5
.PRINT DC V(1) V(2)
.END
```

LN( ) Subcircuit

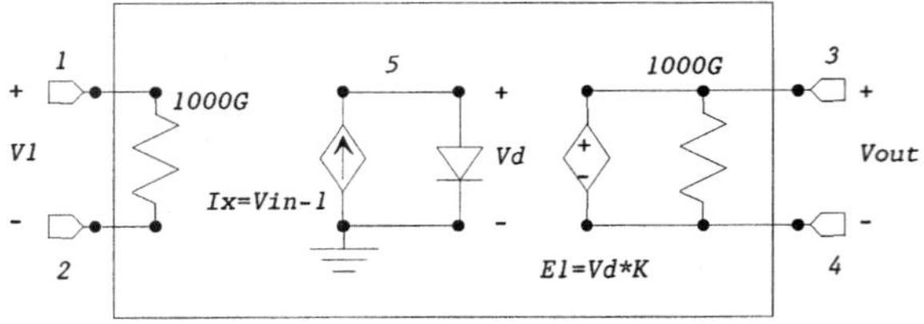

Sine Subcircuit

```
Advanced Techniques
*
* RCG Research, Inc. 3/13/90
*
* This is a two port sine wave circuit.
* The output voltage is sin(vin)
*
.SUBCKT SIN 1 2 3 4
*
* 1: Vin+
* 2: Vin-
* 3: Vout+
* 4: Vout-
*
R12 1 2 1000G
R34 3 4 1000G
ROUT 3 4 1000G
E1 3 4 POLY(1) (1,2) 0 1 0 -.16667 0 8.33333M 0 -198.413U 0 2.75573U
+0 -25.0521N 0 160.590P 0 764.716F 0 -2.81146F 0 .00822064F
.ENDS
*
* Test the circuit with a voltage sweep
*
V1 1 0 DC 1
X1 1 0 2 0 SIN
R1 2 0 1000G
.DC V1 -3.14 3.14 10M
.PRINT DC V(1) V(2)
.OPTIONS LIMPTS=1MEG
.END
```

Sine Wave Subcircuit

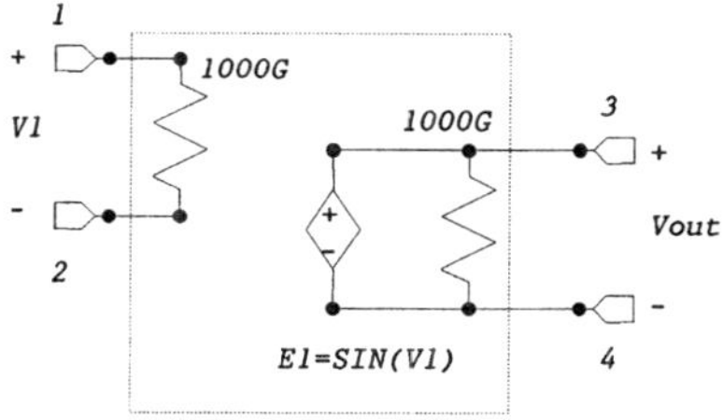

Cosine Subcircuit

```
Advanced Techniques
*
* RCG Research, Inc. 3/13/90
*
* This is a two port cosine wave circuit.
* The output voltage is cos(vin)
*
.SUBCKT COS 1 2 3 4
*
* 1: Vin+
* 2: Vin-
* 3: Vout+
* 4: Vout-
*
R12 1 2 1000G
R34 3 4 1000G
ROUT 3 4 1000G
E1 3 4 POLY(1) (1,2) 1 0 -.5 0 41.66666M 0 -1.38888M 0 24.8016U 0 -275.573N
+0 2.08768N 0 -11.4707P 0 47.7948F 0 -.156192F
.ENDS
*
* Test the circuit with a voltage sweep
*
V1 1 0 DC 1
X1 1 0 2 0 COS
R1 2 0 1000G
.DC V1 -3.14 3.14 10M
.PRINT DC V(1) V(2)
.OPTIONS LIMPTS=1MEG
.END
```

Cosine Wave Subcircuit

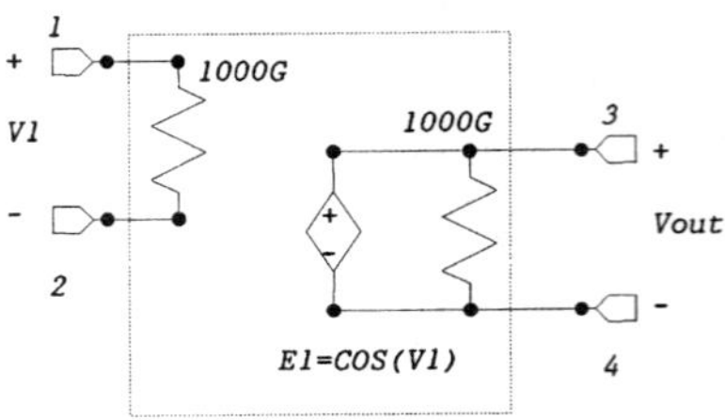

Tangent Subcircuit

```
Advanced Techniques
*
* RCG Research, Inc. 3/13/90
*
* This is a two port sine wave circuit.
* The output voltage is sin(vin)
*
.SUBCKT TAN 1 2 3 4
*
* 1: Vin+
* 2: Vin-
* 3: Vout+
* 4: Vout-
*
X1 1 2 5 6 SIN
X2 1 2 7 8 COS
X3 5 6 7 8 3 4 DIV
RP1 5 0 1000G
RP2 7 0 1000G
.INCLUDE SIN.MOD
.INCLUDE COS.MOD
.INCLUDE DIV.MOD
.ENDS TAN
*
* Test the circuit
*
V1 1 0 DC 1
X1 1 0 2 0 TAN
R1 2 0 1000G
.DC V1 -3.14 3.14 10M
.PRINT DC V(1) V(2)
.OPTIONS LIMPTS=1MEG
.END
```

Tangent Wave Subcircuit

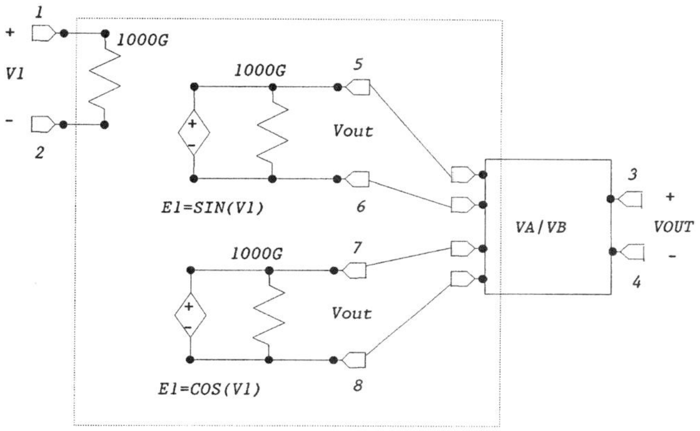

Voltage-Controlled Resistor

```
.SUBCKT VRES 10 20 60 70
*
* RCG Research, Inc. 1/29/92
*
* This is a (linear) voltage controlled resistor
* The resistance is equal to Vc*Rin
*
* 10: R+
* 20: R-
* 60: VC+
* 70: VC-
*
RIN 30 20 1K
VIN 10 30 DC=0
FIN 20 30 POLY(2) VIN VI 0 0 0 0 1
* CONTROL VOLTAGE
RC1 60 70 1G
RC2 60 70 1G
RC3 70 0 1G
* INTERMEDIATE CONTROL STAGE
EI 40 0 POLY(1) (60,70) -1 1
RI 50 0 1
VI 40 50 DC=0
.ENDS VRES
```

Linear Voltage-Controlled Resistor

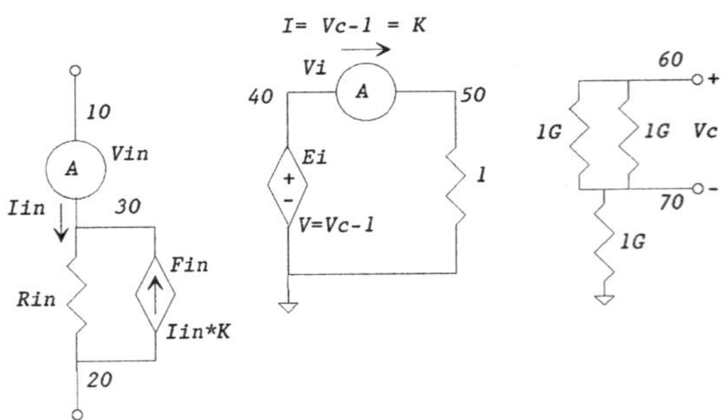

Voltage-Controlled Capacitor

```
.SUBCKT VCAP 10 20 60 70
*
* RCG Research, Inc. 1/29/92
*
* This is a (linear) voltage controlled capacitor
* The capacitance is equal to Vc*Cin
*
* 10: C+
* 20: C-
* 60: VC+
* 70: VC-
*
CIN 30 20 1UF
VIN 10 30 DC=0
FIN 10 20 POLY(2) VIN VI 0 0 0 0 1
* CONTROL VOLTAGE
RC1 60 70 1G
RC2 60 70 1G
RC3 70 0 1G
* INTERMEDIATE CONTROL STAGE
EI 40 0 POLY(1) (60,70) -1 1
RI 50 0 1
VI 40 50 DC=0
.ENDS VCAP
```

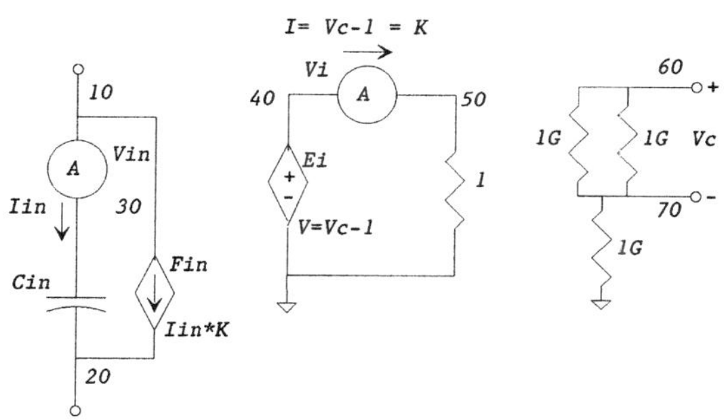

Voltage-Controlled Inductor

```
.SUBCKT VIND 10 20 60 70
*
* RCG Research, Inc. 1/29/92
*
* This is a (linear) voltage controlled capacitor
* The capacitance is equal to Vc*Lin
*
* 10: L+
* 20: L-
* 60: VC+
* 70: VC-
*
LIN 30 20 1UH
EIN 10 30 POLY(2) (30,20) (50,0) 0 0 0 0 1
* CONTROL VOLTAGE
RC1 60 70 1G
RC2 60 70 1G
RC3 70 0 1G
* INTERMEDIATE CONTROL STAGE
EI 40 0 POLY(1) (60,70) -1 1
RI 50 0 1
VI 40 50 DC=0
.ENDS VIND
```

Linear Voltage-Controlled Inductor

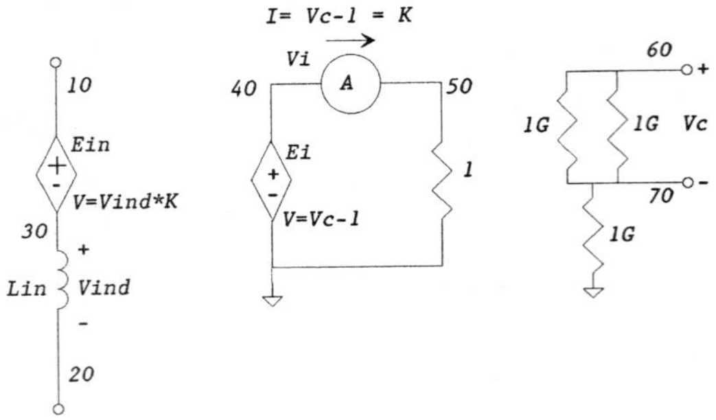

Voltage-Controlled Potentiometer

```
.SUBCKT VPOT 10 40 20 80 90
*
* RCG Research, Inc. 1/29/92
*
* This is a (linear) voltage controlled potentiometer
* The resistance of the pot is equal to R1+R2
* The control voltage may vary from  1 : Vout = Vin
*                                  to  .5 : Vout = Vin/2
*                                  to   0 : Vout = 0
*
* NOTE: R1 must equal R2 for the pot to work correctly!
*
* 10: V+
* 40: Vout
* 20: V-
* 80: Vcontrol +
* 90: Vcontrol -
*
R1 30 40 1K
R2 40 20 1K
FR1 40 30 POLY(2) VIN VEC1 0 0 0 0 1
FR2 40 20 POLY(2) VIN VEC1 0 0 0 0 1
VIN 10 30 DC=0
* Control voltage
RC1 80 90 1000G
RC2 80 90 1000G
RC3 90  0 1000G
* First intermediate stage
EC1 50 0 (70,0) -1
REC1 60 0 1
VEC1 50 60 DC=0
* Second intermediate stage
EC2 70 0 POLY(1) (80,90) -1 2
REC2 70 0 1
.ENDS VPOT
```

Linear Voltage-Controlled Potentiometer

All models used in this Appendix were reproduced from “Successfully Simulating Circuits with SPICE.” Copyright RCG Research, Inc. Used with permission.

References

1. *RSPICE User's Guide*, RCG Research, Inc., 1990.

Appendix

E

Using the RS_CALC Program

RS_CALC is a FORTRAN program which is used to calculate the source and drain series resistance of a JFET or MOSFET transistor. To run the RS_CALC program, change into the UTIL subdirectory (after installing the software). Enter RS_CALC at the DOS prompt.

```
RS_CALC
```

The program will respond with:

```
Enter VDS_>
```

Enter the value of V_{ds} used for the linear characteristic test. The program will continue with:

```
Enter Vgs_>
```

This value is normally zero for a depletion mode device (such as a JFET) and nonzero for an enhancement mode transistor. The program will continue with:

```
Enter the calculated VTO_>
```

Enter the value of VTO calculated from the linear characteristics. The program will continue with:

```
Enter the calculated BETA (for JFETs)
or KP/2 (for MOSFETs)_>
```

Enter the value of BETA for the JFET model or half the value of KP for a MOSFET. The value of BETA and KP are obtained from the linear characteristics of the transistors. These values must be calculated prior to running this program. The program will continue with:

```
Enter Id(high)_>
```

Enter the higher (ideal) value of I_d. This represents the value of I_d if the slope of the low I_d curve was extrapolated to the V_{gs} voltage. See Fig. E.1 for a graphic illustration of $I_{d(high)}$ and $I_{d(low)}$ for (*a*) N-channel MOSFETs and (*b*) N-channel JFETs. The program will continue with:

```
Enter Id(low)
```

Enter the lower value of I_d. This represents the actual measured value of I_d at V_{gs}.

The program will then calculate RS and RD, and the calculated value of RS and RD will be displayed on the screen and stored in the file named RS_CALC.OUT. For most JFETs and MOSFETs, the drain and source resistance will be equal and, because of this, the program assumes RS and RD are symmetrical and equal.

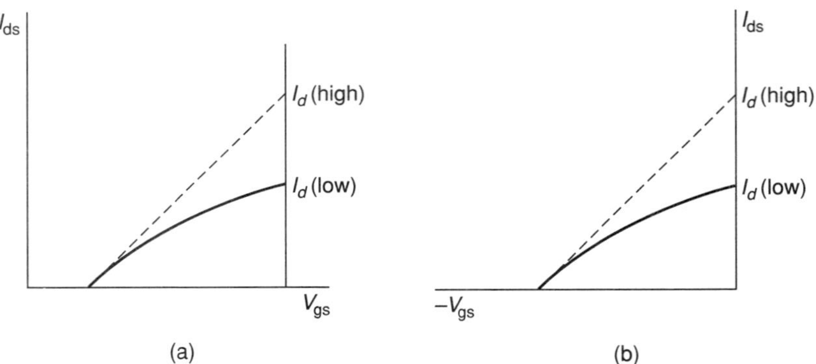

Figure E.1 (*a*) Linear MOSFET characteristics and (*b*) linear JFET characteristics.

Index

ABOUT THE AUTHOR

Ron Kielkowski developed the *Successfully Simulating Circuits with SPICE* and the *MicroSim PSpice Workshop* training seminars for RCG Research. He is the primary instructor for these courses and currently teaches them throughout the United States. Mr. Kielkowski also led the development of the RSPICE and RGRAPH programs for RCG Research and is the author of the text *Inside SPICE*. Previously, Mr. Kielkowski directed simulation activities at both the IC Design Center of Delco Electronics and the Advanced Development Center of the Naval Air Warfare Center.

ABOUT THE DISK

On the disk which accompanies this text are the RSPICE and RGRAPH programs. RSPICE is a PC version of Berkeley SPICE2G.6. RGRAPH is a sophisticated graphical postprocessor which may be used to view and plot simulation results. Both of these programs use extended memory and true 32-bit adressing. This makes these programs exceptionally fast and powerful.

RSPICE and RGRAPH were written by RCG Research, Inc. for use with their *Successfully Simulating Circuits with SPICE* training classes. Since much of the material for this text comes from these same classes, it seemed appropriate to include the programs as well. Many of the graphics illustrated in the text were generated with RSPICE and RGRAPH.

The RSPICE and RGRAPH programs provided with this text are free programs with a reserved copyright. Individuals may use, copy, and distribute these programs free of charge. Users may wish to register their copies with RCG Research for technical support and upgrade notices as the programs evolve.

The program disk also contains the utility programs described in Chaps. 2, 3, 4, 5, and 6. The utility programs aid the individual in extracting model parameters, temperature coefficients, and element values. Extracting model parameters without these utility programs can be done, but the process is extremely tedious. These utility programs will be an invaluable aid for anyone interested in developing device models for SPICE.

A library of analog behavioral models is also included on the disk. Analog behavioral models allow an individual to build SPICE models by describing voltages and currents in terms of mathematical relationships (rather than the normal electrical relationships). To describe the temperature effects of resistors, capacitors, and inductors, analog behavioral models will be used. A power MOSFET model will be constructed with behavioral elements.

Eighteen different behavioral building blocks have been included on the disk for the user. With these building blocks, other behavior models can be developed as needed.

Installing the Disk

The accompanying RSPICE/RGRAPH disk contains the RSPICE circuit simulator and the RGRAPH graphical postprocessor. The disk also contains the circuit examples used throughout this text. An installation program automates the task of creating the proper directory structure, copying the files, and assigning the proper directory pointers.

Requirements

To install the programs and files you will need approximately 3MB of free space on your hard disk. The install program creates the following directories and subdirectories.

```
\RCGV34
\RCGV34\RSPICE
\RCGV34\CH2
\RCGV34\CH3
\RCGV34\CH4
\RCGV34\CH5
\RCGV34\CH6
\RCGV34\UTIL
\RCGV34\LIB
```

The RSPICE and RGRAPH files are then copied to the appropriate directories.

Installation

We'll assume you're installing the software from a floppy disk. To install the programs, simply type:

```
X:install X: Y:
```

where X is the floppy disk drive designator and Y is the hard disk drive designator. The installation program will complete the remaining installation chores.

SOFTWARE AND INFORMATION LICENSE

The software and information on this diskette (collectively referred to as the "Product") are the property of McGraw-Hill, Inc. ("McGraw-Hill") and are protected by both United States copyright law and international copyright treaty provision. You must treat this Product just like a book, except that you may copy it into a computer to be used and you may make archival copies of the Products for the sole purpose of backing up our software and protecting your investment from loss.

By saying "just like a book," McGraw-Hill means, for example, that the Product may be used by any number of people and may be freely moved from one computer location to another, so long as there is no possibility of the Product (or any part of the Product) being used at one location or on one computer while it is being used at another. Just as a book cannot be read by two different people in two different places at the same time, neither can the Product be used by two different people in two different places at the same time (unless, of course, McGraw-Hill's rights are being violated).

McGraw-Hill reserves the right to alter or modify the contents of the Product at any time.

This agreement is effective until terminated. The Agreement will terminate automatically without notice if you fail to comply with any provisions of this Agreement. In the event of termination by reason of your breach, you will destroy or erase all copies of the Product installed on any computer system or made for backup purposes and shall expunge the Product from your data storage facilities.

LIMITED WARRANTY

McGraw-Hill warrants the physical diskette(s) enclosed herein to be free of defects in materials and workmanship for a period of sixty days from the purchase date. If McGraw-Hill receives written notification within the warranty period of defects in materials or workmanship, and such notification is determined by McGraw-Hill to be correct, McGraw-Hill will replace the defective diskette(s). Send request to:

Customer Service
McGraw-Hill, Inc.
Gahanna Industrial Park
860 Taylor Station Road
Blacklick, OH 43004-9615

The entire and exclusive liability and remedy for breach of this Limited Warranty shall be limited to replacement of defective diskette(s) and shall not include or extend to any claim for or right to cover any other damages, including but not limited to, loss of profit, data, or use of the software, or special, incidental, or consequential damages or other similar claims, even if McGraw-Hill has been specifically advised as to the possibility of such damages. In no event will McGraw-Hill's liability for any damages to you or any other person ever exceed the lower of suggested list price or actual price paid for the license to use the Product, regardless of any form of the claim.

McGRAW-HILL, INC. SPECIFICALLY DISCLAIMS ALL OTHER WARRANTIES, EXPRESS OR IMPLIED, INCLUDING BUT NOT LIMITED TO, ANY IMPLIED WARRANTY OF MERCHANTABILITY OR FITNESS FOR A PARTICULAR PURPOSE. Specifically, McGraw-Hill makes no representation or warranty that the Product is fit for any particular purpose and any implied warranty of merchantability is limited to the sixty day duration of the Limited Warranty covering the physical diskette(s) only (and not the software or information) and is otherwise expressly and specifically disclaimed.

This Limited Warranty gives you specific legal rights; you may have others which may vary from state to state. Some states do not allow the exclusion of incidental or consequential damages, or the limitation on how long an implied warranty lasts, so some of the above may not apply to you.

This Agreement constitutes the entire agreement between the parties relating to use of the Product. The terms of any purchase order shall have no effect on the terms of this Agreement. Failure of McGraw-Hill to insist at any time on strict compliance with this Agreement shall not constitute a waiver of any rights under this Agreement. This Agreement shall be construed and governed in accordance with the laws of New York. If any provision of this Agreement is held to be contrary to law, that provision will be enforced to the maximum extent permissible and the remaining provisions will remain in force and effect.